A Comprehensive Guide to Project Management Schedule and Cost Control

A Comprehensive Guide to Project Management Schedule and Cost Control

Methods and Models for Managing the Project Lifecycle

Randal Wilson

Associate Publisher: Amy Neidlinger
Executive Editor: Jeanne Glasser Levine
Operations Specialist: Jodi Kemper
Cover Designer: Chuti Prasertsith
Managing Editor: Kristy Hart
Project Editor: Andy Beaster
Copy Editor: Chuck Hutchinson
Proofreader: Paula Lowell
Indexer: Cheryl Lenser
Senior Compositor: Gloria Schurick
Manufacturing Buyer: Dan Uhrig

© 2014 by Randal Wilson
Publishing as Pearson
Upper Saddle River, New Jersey 07458

For information about buying this title in bulk quantities, or for special sales opportunities (which may include electronic versions; custom cover designs; and content particular to your business, training goals, marketing focus, or branding interests), please contact our corporate sales department at corpsales@pearsoned.com or (800) 382-3419.

For government sales inquiries, please contact governmentsales@pearsoned.com.

For questions about sales outside the U.S., please contact international@pearsoned.com.

Company and product names mentioned herein are the trademarks or registered trademarks of their respective owners.

All rights reserved. No part of this book may be reproduced, in any form or by any means, without permission in writing from the publisher.

Printed in the United States of America

First Printing April 2014

ISBN-10: 0-13-357294-3
ISBN-13: 978-0-13-357294-0

Pearson Education LTD.
Pearson Education Australia PTY, Limited.
Pearson Education Singapore, Pte. Ltd.
Pearson Education Asia, Ltd.
Pearson Education Canada, Ltd.
Pearson Educación de Mexico, S.A. de C.V.
Pearson Education—Japan
Pearson Education Malaysia, Pte. Ltd.

Library of Congress Control Number: 2014930853

*I would like to dedicate this book to my wife, Dusty,
and sons, Nolan, Garrett, and Carlin,
for their continued support and patience on this project.*

Contents

Dedication v
About the Author x
Introduction 1
    Schedule and Cost of Projects 1
    Project Balance 2
    What Is Control? 3
    Organizational Influences 4
    Solutions to Schedule and Cost Control 9

Part 1	**Project Development** 11
Chapter 1	**Basic Project Structure** 13
	Introduction 13
	Projects, Programs, and Portfolios 14
	Project Management Versus Program and Portfolio Management 17
	Project Life Cycle 23
	Review Questions 25
Chapter 2	**Initiating Process** 27
	Introduction 27
	Project Origination 28
	Project Stakeholders 31
	Project Selection 34
	Project Charter 52
	Review Questions 55
Chapter 3	**Planning Process** 57
	Introduction 57
	Develop Project Management Plan 58
	Collect Requirements 63
	Define Scope 66
	Work Breakdown Structure (WBS) 70
	Review Questions 75
Part 2	**Project Schedule Analysis** 77
Chapter 4	**Activity Definition** 79
	Introduction 79
	Activity Analysis 80
	Responsibility Assignment 88

		Work Authorization 91
		Review Questions 95
		Applications Exercise for Chapters 4, 5, 6, 7, and 8 95
		Case Study Exercise for Chapter 4 96

Chapter 5 Activity Sequencing 97
- Introduction 97
- Information for Sequencing 98
- Defining Dependencies 102
- Precedence Diagramming Method (PDM) 104
- Review Questions 116
- Applications Exercise 116

Chapter 6 Resource Estimating 117
- Introduction 117
- Types of Resources 119
- Resource Constraints 122
- Resource Requirements 126
- Resource-Estimating Methods 128
- Review Questions 136
- Applications Exercise 137

Chapter 7 Activity Duration Estimating 139
- Introduction 139
- Duration Estimating Methods 141
- Duration Estimating with Constraints 146
- Scheduling Conclusions 151
- Review Questions 155
- Applications Exercise 156

Chapter 8 Schedule Development 157
- Introduction 157
- Schedule Requirements 158
- Schedule Structuring Techniques 162
- Schedule Analysis 171
- Schedule Documentation 180
- Review Questions 182
- Applications Exercise 182

Part 3 Project Cost Analysis 183

Chapter 9 Cost Estimating 185
- Introduction 185
- Collecting Cost Data 186

	Cost Constraints	191
	Estimating Tools and Techniques	192
	Review Questions	198
	Applications Exercise for Chapters 9 and 10	198
Chapter 10	**Budget Development**	**201**
	Introduction	201
	Functions of a Budget	202
	Budget Development Methods	205
	Budget Constraints	208
	Cost of Quality	210
	Review Questions	212
	Applications Exercise	213
Part 4	**Project Monitoring and Control**	**215**
Chapter 11	**Schedule and Cost Monitoring**	**217**
	Introduction	217
	Integrated Monitoring	218
	Monitoring and Analysis Tools	221
	Troubleshooting Tools	233
	Monitoring Results	235
	Review Questions	239
	Applications Exercise	239
	Case Study Exercise for Chapter 11	240
Chapter 12	**Schedule and Cost Control**	**241**
	Introduction	241
	Change Control	242
	Control Tools and Techniques	247
	Control Results	260
	Review Questions	263
	Applications Exercise	263
	Bibliography	**265**
	Index	**267**

About the Author

Randal Wilson, MBA, PMP, serves as Visiting Professor of Project Management, Keller Graduate School of Management, at the Elk Grove, California, DeVry University campus. His teaching style is one of addressing project management concepts using academic course guidelines and text and includes in-depth discussions in lectures using practical application from industry experience.

Mr. Wilson is currently Operations and Project Manager at Parker Hose and Fittings. He is responsible for five locations across northern California and Nevada, as well as project management of redesigns and renovation of existing facilities and construction of new facilities.

Mr. Wilson was formally in the telecommunications industry as Senior New Product Introduction Engineer at REMEC, Inc., Senior New Product Introduction Engineer with Spectrian Corp., and Associate Design Engineer with American Microwave Technology. He also served as Senior Manufacturing Engineer at Hewlett-Packard.

He is a certified Project Management Professional (PMP) of the Project Management Institute. He acquired an MBA with a concentration in General Operations Management from Keller Graduate School of Management of DeVry University in Fremont, California, and a bachelor of science in Technical Management with a concentration in Project Management from DeVry University in Fremont, California.

Introduction

Schedule and Cost of Projects

Most organizations are formed for a purpose that results in producing goods or services. The success of the organization's endeavor is in the management of resources and how the founders of the organization have structured the operation to achieve its strategic objective. Because most organizations require resources to facilitate the ability to accomplish daily tasks in the operation, on some occasions special activities are required to accomplish certain things the operation needs that are not part of the daily tasks, but require resources from within the organization. These types of special tasks are called *projects,* and if they are structured and managed well, they will provide opportunities for the organization to make improvements that are necessary in the ongoing improvement of the operation.

Depending on the type and size of the organization, projects may be sporadic and used only in special development situations, whereas other organizations may use projects integrated into its business structure as a main part of its daily operation. Regardless of how projects are utilized in the organization, they require resources that may include human resources, equipment and materials, facilities, and financial resources. When special projects require these types of resources, it is important to note that most of these resources are utilized in normal daily operation tasks; if they are used on a special project, they have to be allocated such that they do not impact daily operations and create conflicts. Most resources used within an organization have a cost component associated with how they are used; how a resource is expensed for a special project also is a consideration.

Organizations can generally benefit from special projects, but the structure, organization, and utilization of resources become a very important element not only in the success of the project, but in minimizing the impact to the organization and daily operations. Before a special project can be authorized, management within the organization needs to know how much the project will cost, how much of the organization's resources will be required, what are the expected deliverables or benefit to the organization, and how long the project will take to complete. Because it is usually easier to identify how a project deliverable will benefit the organization, it can be difficult to ascertain how much the project will cost, how long it will take, and how many and what types of resources will be required to complete the project objective. At this point, project management tools and techniques can be utilized to define cost and schedule requirements.

Project Balance

When an organization embarks on a special project, generally, the first task is to define what the project objective will be as it relates to expected deliverables. This task is usually accomplished fairly quickly by initial stakeholders; they include operations managers and other staff interested in the benefit the project deliverables will bring to the operation. Problems generally begin to arise when the project has been approved and the process of defining how to create the project deliverable, what costs will be associated, what resources will be required, and how to schedule these resources for a special project in addition to their normal daily tasks must be considered. Projects have to be structured and managed, maintaining a balance between utilizing the organization's resources for daily operations and for required tasks to complete work activities on a project.

Resources can be human resources and any other resources, including financial requirements, that will be needed on a project; however, they have to be defined, showing how they can be utilized in the balance of daily operations tasks and special project tasks. This can create a challenge for operations managers because they have an obligation to daily work activities but want to see special projects completed and struggle with ways to balance human resources and other resources within their department to complete both of these tasks. Part of this dilemma stems from the department manager being loyal to his obligation in the daily operations and will typically err to ensuring operations are not impacted, thus creating constraints for utilizing resources on special projects. The other part of this dilemma might be the department manager's inexperience with scheduling resources for both departmental tasks and project tasks. Project managers who are separated from operational departments do not share the same loyalty and, in fact, are probably more loyal to completing project tasks and will negotiate with departmental managers in scheduling resources for project tasks. Project managers can also solicit the use of resources from outside the organization to complete project activities if internal resources are simply not available.

The second component that presents a challenge with special projects in an organization is defining all the costs associated with project activities and developing an overall budget for a special project. Typically, the error in costing projects is looking at project activities at a high level and trying to associate a more "generalized" cost to complete the entire activity, not taking into account all the specifics within each activity. Although this approach can be used in cost estimating a project, it is typically the reason that projects go over budget and that project activity costs are difficult to control. If project activity costs have been established based on the overall activity, as the activity plays out and specific tasks have resource, material, and equipment requirements that have not been accounted for, these requirements can add costs. Because these resources are needed, these costs present challenges in trying to control costs for the overall activity. Project managers typically have more time and techniques to break down activities to understand all the associated costs to derive a more accurate work activity cost and project budget.

Organizations that do not have project management capabilities will ultimately struggle with the implementation of projects in defining project costs, scheduling resources, and controlling project work activities. The key element, in addition to properly costing and scheduling a project, is in the control of costs and scheduling of activities and resources to ensure a project completes an objective at the estimated cost and within the allotted schedule.

What Is Control?

Projects can be developed and managed within an organization under the direction of the department manager for the sole purpose of completing a unique objective for that department. Projects also can be developed in an environment where several resources throughout the organization can be used to complete project activities. Either a department "functional" manager or a project manager can manage projects. These projects experience similar project life-cycle phases. One aspect of projects is consistent no matter what type of organizational structure or how big the project is: Projects have costs and schedules and need oversight and adjustments made to keep project activities within budget and on schedule. This is called *project control*.

Reporting Versus Managing

Overseeing project activities puts managers in a position of responsibility to ensure that project activities are completed. How managers view their responsibility plays a large role in whether the project is controlled or simply monitored. Managers will find themselves in one of two managerial roles with regards to projects: (1) monitoring and reporting activities; or (2) assigning, monitoring, and controlling activities. When managers simply report the status of project activities, this is not a control function. It is simply an observation of what is happening and reporting of status. *Control* in project management is defined as *having a means of measurement and initiating adjustments in the course of an activity to address unwanted changes to cost, schedule, quality, or risk elements that have influenced the activity.*

The Manager's Role in Control

Project managers are educated and/or trained in the need to provide control within the activities of a project. This requires the project managers' active participation in not only monitoring activities against a baseline of estimated cost and schedule, but also initiating adjustments that bring activities back in line with budget and schedule if problems arise. Either functional or project managers can achieve control over a project as long as they understand what control is designed to do for activities within a project. Control of the project is one of the most important roles project managers can have with oversight of project activities. One might say that anyone can observe project activities and report on status, but real management of a project has an element of control such that actively adjusting activities results in improvements to cost or schedule. Inasmuch as project managers utilize tools and techniques not only to monitor but also to control project activities, other forces and influences within the organization can present challenges to the success of a project. Project managers and/or functional managers must be aware of influences unique to the organization that can impose restrictions, constraints, and even conflicts for special projects operating within an organization.

Organizational Influences

Projects can play an important role in the success of an organization, but the development and management of these structures alone will not result in isolated entities within the organization. However, these roles are still subject to other internal and external influences that can make or break the goal of completing objectives. Projects simply give the organization focus and the ability to control activities required to complete special objectives within the organization. Because the organization typically has established departments to complete certain activities for daily operations, some of these areas produce things for profit, called profit centers; other areas within the operation complete tasks to support the profit centers, such as administration, accounting, and human resources. Because special projects can utilize resources throughout the organization primarily from within the profit centers, projects are connected to other areas within the organization not associated with profit centers to facilitate completion of strategic objectives. Although these areas are needed, they can present either positive or negative influences on the success of completing projects; therefore, project managers should take them into consideration. Three primary areas within the organization can have a significant influence on how projects are structured, scheduled, budgeted, and controlled, and they have to do with the organization's leadership, culture, and structure.

Organizational Leadership

There is a consistent rule within most organizations that everything starts from the top and rolls down. This rule also is true in the area of managing projects. Whether it is perception or actual fact, the impact this rule will have on an organization starts with the general maturity of the organization and senior staff as well as specific management styles of those overseeing projects. If the executive staff does not understand the importance and benefits of projects, they will not always be supportive of what managers are trying to accomplish and the approach they are taking in using projects to manage activities within the organization. This can come across in several forms, behaviors, attitudes, and actions such as

- Poor selection of key managers in critical roles
- Approval or nonapproval of certain projects and activities
- Unnecessary timelines or budget constraints creating undue stress on projects and activities
- Misunderstanding or ignorance of critical activity update information
- Personality conflicts with project managers
- Hidden agendas that drive inconsistent or confusing decisions

It is important that executive management understand their role in leading by example. They also must understand the impact their leadership can have on the organization if it is not performed at the highest level of integrity, professionalism, and cooperation among themselves and with those reporting to them. It is also important that they understand their actions are seen not only by those reporting to them but by many in the organization; and their leadership can be a large part of the culture established within the organization.

Organizational Culture

When having discussions about the culture of organizations, people can go in several directions to assess, label, and/or stereotype organizations for a perceived culture. When we talk about culture, the general idea is not only the DNA makeup of how the organization structures itself, but also its management style and personality. It is interesting that an organization, in many ways, has a reputation or is known in the industry by its personality and how it conducts business. Some of this personality and management style are a direct result of those who started the organization or are currently senior officers within the organization, whereas other traits of organizational personality might be a result of how the organization conducts its business based on market demands and customer relations. Because these areas are typically seen as high level and generally broad-based perceptions or interpretations of business operations, the same DNA is found at the department and project levels.

It is important that project managers understand the DNA or personality of the organization in the form of a management style so that they can be consistent with the way the organization conducts business internally and externally. This helps project managers be consistent in their management style with the general culture of the organization and can make it easier to gain the approval of senior management. DNA is a complex strand of several elements, and the organization is similar because it is made up of several areas that ultimately define its personality and culture. Some of these areas include

- Type of business and market position
- Senior management experience, personality, and management style
- Hierarchical command structure
- Maturity in customer and supplier relationships
- High-level investment strategies and risk tolerance
- Senior management's perception of lower-level workforces
- Organizational approach to customer service
- General working conditions and environment within the organization

Understanding what makes up the DNA and personality of an organization can help project managers not only understand their place in the organization, but also understand the importance of a successful management style that is in sync with the culture of the organization. This also allows for managers to be more consistent with other peer management styles. The project managers can also benefit in better understanding the mindset and possible perceptions of the workforce, which can help in the project managers' management style and approach with their staff. One element of the organization's DNA is in the type of organization and how it is structured functionally based on the type of business it conducts. The type of management structure used can play a large role in defining how the organization conducts business, its relationships with customers, and the general role project managers will ultimately have.

Organizational Structures

Organizational structure is the foundation of how business is conducted both internally and externally. It plays a large role in how daily operations are carried out and how projects are integrated within daily operations. Some organizations utilize projects at a very low level, accomplishing small tasks, whereas other organizations utilize projects, and their main course of business in the organization is structured with emphasis on these large projects. Depending on how organizations utilize projects within daily operations, organizations are structured using one of three basic structures. These structures are called *functional, projectized,* and *matrix*.

Functional organizations employ the classic structure used to establish managerial hierarchy with the organization divided into traditional functional departments. These departments can include accounting, human resources, purchasing, engineering, manufacturing, quality control, inventory, and warehousing, as well as shipping and receiving. The general idea with this structure is each department has a specific objective with a clear chain of command wherein each department has a manager overseeing the work activities of that department. The manager of each department reports to a higher-level manager who may be overseeing several departments, and the chain of command continues all the way up to the highest level of management in the structure. Although organizations have found this structure to be successful in the general operation of business, it has inherent strengths and weaknesses with regard to efficiency, accountability, and resource management, as well as the management of projects and the role of the project manager.

The main strength of functional organizations is each department performing its activities as a unit and requiring little or no direct involvement with other departments to achieve its objectives. Each department's strength builds on the collective knowledge and experience of its members and processes it has developed to maximize the efficiency of work activities in completing its normal objectives. Likewise, projects developed within an individual department are most efficient using only its department members and overseen by the department manager.

The weakness of this structure is apparent when the organization selects a functional manager to oversee projects and that person may or may not have the experience of a project manager in structuring projects with regard to cost, schedule, resource management, and control. The project can suffer as a result. If a project manager is used in conjunction with this type of project, the project manager carries little or no authority and acts more like an activity expediter.

Projectized organizations use a completely different type of business structure than that of functional organizations where staff members are grouped into workforces that may include representatives from several traditional departments and are tasked with a unique project objective. This organization only has project groups and very few, if any, functional departments. This type of structure also places a high level of importance on project objectives; therefore, projectized organizations hire project managers to structure and oversee projects. The project manager carries a much higher level of authority with oversight of all resources, budget, and scheduling, and responsibility for completion of the project objective.

Most projectized organizations were originally structured in this form as a result of their business strategic objectives. These objectives are based on groups of activities that result in unique output deliverables. Another big advantage of projectized organizations is the flexibility available in the business strategy. Because this structure emphasizes large projects as its main output, these organizations can respond quickly to changes in market demand, allowing them to be successful in both stable and unstable market environments.

Project management within a projectized organization requires management of activities utilizing different types of resources that can be permanently assigned to the project, borrowed from several departments within the organization, and possibly contracted from resources external to the organization. Unlike a specific project designed to accomplish a goal within a single department, projects are now the goal of the entire organization and may require only a few actual departments such as administration and engineering. Because the organization is structured for projects, human resources are assigned tasks based on the requirements of their skill for specific activities on the project. After they complete their activities, they are reassigned to another project to provide their skills for activity requirements on that project. Human resources in this type of organization spend all their employment moving from project to project.

Matrix organizations are a blend of functional and projectized structures using the benefits of each in completing the organization's objectives. Matrix organizations typically have a combination of routinely produced deliverables as well as unique and specialized projects. This allows for traditional departments led by functional managers to manage output deliverables of their individual departments; the organization also is able to use these same resources in special projects. The functional manager still holds authority over her department, but the project manager can hold an equal level of authority in overseeing resources from several departments in managing a project.

Matrix organizations have the advantage of structure and stability found in functional organizations through established departments. They also use key resources within these departments on projects that allow the organizations the flexibility to produce deliverables in response to changing market conditions. This capability gives senior management a unique opportunity to assess market conditions and in parallel create a stable and predictable product delivery environment and a quick response project environment that are both successful in the marketplace.

Solutions to Schedule and Cost Control

Projects ultimately are the utilization of organizational resources identified for specific tasks that have been organized in a sequence of work activities that will accomplish the project objective. Although most organizations have management staff who are very good at utilizing resources for specific tasks within the operation, the trick in *project management* is not only to specifically identify all required resources, correctly sequence work activities, and accurately estimate costs of all specific work activity requirements, but also to design and initiate schedule, cost, and quality controls. As functional managers can probably accomplish some of these project-related tasks, their responsibility is not in directing daily work activities of the department. As we have seen, if a project is unique to a particular department, functional managers can actually be the best people to carry out that specific project as the scope management will be isolated to their department. If projects require resources from multiple departments or projects occupy a primary structure within the organization, project managers typically have the knowledge and experience to accurately develop and implement projects.

If project managers want to ensure that a project is developed properly, they must be knowledgeable of tools and techniques that will assist in correctly and accurately gathering and evaluating project information to develop a comprehensive project management plan. The success of projects is typically the result of having well-documented project work activity requirements, accurate scheduling of resources and work activities, and accurate cost estimation of all specific activity requirements to develop the project budget. One of the primary success factors in project management is the attention to detail in work activity requirements and scheduling.

The goal of successfully developing and controlling a project would be understanding how to get the most detail of work activities to accurately assess cost and resource requirements and what types of controls can be implemented to ensure work activities stay on schedule and within budget. In most cases, the more accurate the project manager and project staff are in estimating schedule duration and cost for work activities, the better chance the project will have staying on schedule and within budget. Because there typically is a margin of error in schedule and cost estimating and the reality of the impact of organizational processes and risk or uncertainty, the project manager also needs to utilize some control to adjust for these abnormalities in the project life cycle.

The success of the project ultimately depends on how well the project manager is armed with tools and techniques for project development. This book is a comprehensive compilation of common project management tools and techniques used in project development, specifically with regard to scheduling, cost estimating, and project control. These tools are simple and can be used easily in the development, implementation, and control of a project. It should be the goal of all those tasked with the development and oversight of a project to be armed with tools that will assist in projects being documented and controlled effectively and accurately to ensure project objectives are completed on schedule and on budget.

Part 1

Project Development

Chapter 1   Basic Project Structure 13

Chapter 2   Initiating Process 27

Chapter 3   Planning Process. 57

Organizations large and small around the world find it necessary to engage in endeavors that will produce a unique deliverable. This task, called *project management*, requires gathering and organizing information work activity in the management of resources to complete project activities. When an organization determines that there is a need or has been informed of a requirement, the initial process in developing a project begins with identification of stakeholders in the gathering of information to decide whether the organization should officially engage in the endeavor.

The chapters in Part 1 detail the initial processes required to authorize a project and begin the initial planning stages. The organization should take seriously all potential projects, whether large or small, and the level of management required for the potential scope of the projects. Most projects follow a similar process, as outlined in these chapters. Those responsible for deciding whether a project should begin must understand the scope of the project. It is important that organizations follow a process for the evaluation of potential projects to ensure critical components of information are gathered and analyzed for the decision-making process. It is also important that organizations follow a process to be consistent in the way decisions are made for potential projects and that the appropriate stakeholders have been identified to make these decisions on behalf of the organization.

It is the ultimate goal of the project manager to develop an overall project management plan that defines four primary functions the project manager will carry out in the project life cycle:

- Develop a master schedule of project activities
- Develop a proposed project budget
- Monitor and report the status of project activities
- Implement controls to ensure project activities stay on schedule and budget

1

Basic Project Structure

Introduction

Organizations are *formed for a purpose,* usually resulting in the development and production of a product or the offering of a service. Because both of these types of organizations incur costs to produce products and services, these costs must be maintained to ensure the profitability of these organizations. How costs and schedules are not only estimated but also controlled is largely a function of how an organization is structured and the role project managers might have within the organization.

It is imperative that organizations structure themselves in a way that will allow costs of all departments to be measured, analyzed, controlled, and maintained. This structure is the first level of cost control, as it separates the organization into cost centers and helps define what these departments are, how big they need to be, and what purpose they serve within the organization. Organizational structuring has been common for decades; when organizations are structured correctly, they have been proven more successful in monitoring and controlling their costs. This may sound like Business 101, but running a project is much like running a business, and a well-structured and organized project is one that is easier to control.

The project manager should approach structuring a project much like dividing an organization into groups and within each group breaking down smaller units until all finite costs can be identified and measured. When projects are categorized and broken up into smaller subcomponents, task scheduling and cost estimating are much easier to identify, analyze, and control. Many organizations divide work into projects or programs and therefore group project costs based on output or deliverables. Organizations can monitor their cost centers and develop strategic planning based on the success of various projects or programs. The project manager can also monitor costs and scheduling of resources on a project and manage strategic adjustments based on real-time performance compared to what was originally planned. To accomplish this, the project manager needs to develop a plan of all activities that need to be accomplished and to include all cost estimate and resource scheduling requirements needed to complete a desired deliverable. To create this plan and develop accurate cost estimating, scheduling, and design of controls, the manager must make sure he understands some basic fundamentals of projects, programs, and portfolio management.

Projects, Programs, and Portfolios

To explain what projects, programs, and portfolios are and how they are used, we must step back and review some basic concepts of organizational structure. As organizations grow in size and complexity, they need to be broken up into functional areas to better define and control the work activities of each area. This helps in managing work being performed, controlling cost expenditures, and maintaining accountability for deliverables required of each area. The work performed in most organizations falls into one of two general categories: work being performed to produce a product or service and work being performed to support the organization.

Producing Versus Supporting

It is important to understand the difference between departments that support an organization versus departments that produce a product or service because this distinction typically defines support versus profit centers and helps clarify whether projects or programs will be used. An example of a support center is the accounting department or human resources department within an organization. These types of departments do not necessarily perform work that creates a profit for the company, but they perform activities that support several areas within the organization. These duties typically are seen as day-to-day functions carried out repetitively or an ongoing basis. Organizations may have several support functions or areas such as warehousing, shipping and receiving, quality control, manufacturing, engineering, and administrative and executive staff.

Profit centers are those areas of the organization that produce output such as a product or service that has an associated cost and will be sold at a higher value to create a profit for the organization. Some organizations produce the same product or service over and over, and there is little or no unique aspect to that product or service. Other organizations might produce a product or service that is more customized and unique and may be produced only in that form one time for a customer.

An example of a unique service or product might be a construction company that builds custom residential homes. Each home is designed and constructed for its unique value to the customer. The construction company performs this action on a regular basis, but the product itself is unique every time it is produced.

Other organizations might produce a manufactured item on an assembly line where the same item is produced in large quantities on a repetitive basis. In this case, the quality of the product is based on the ability to repeat the process the exact same way several times.

The fundamental rule in business is there is an associated cost with the production of a product or service, whether it is one-time and unique, or produced several times on a production line and sold at a target profit margin. Because the sale price of a product or service can be changed based on market conditions, the cost of the product or service must be maintained and/or improved to maintain a profit margin. It is important to understand the difference between one-time unique products or services and a product built several times on a production line because this distinction can help in understanding the basic principles of projects, programs, and portfolios and how they are used in establishing and maintaining costs as well as scheduling resources.

Project Management Structures

Projects, programs, and *portfolios* are common terms that have been used within organizations for many years, but unfortunately all too often they are misunderstood or simply used in the wrong context. The project manager must understand the basic principles behind these terms to help identify how to estimate cost and schedule as well as control a project or program. These three terms identify what type of management structure is required to carry out the scope of activities:

Project—A unique endeavor designed for a specific purpose to accomplish a single objective. Projects have a defined start and end.

Program—A grouping of related projects. Programs can be open-ended with related projects entering and exiting throughout the life of the programs. Programs can have a start but not necessarily an end.

Portfolio—A grouping of either related or nonrelated programs, projects, or individual work activities. Portfolios also can have a start but not necessarily an end.

Based on these definitions, we can conclude that three basic levels of management structure are designed to categorize, prioritize, and group activities within an organization to achieve a strategic objective. Figure 1.1 and Table 1.1 illustrate how the structure of projects, programs, and portfolios works within an organization.

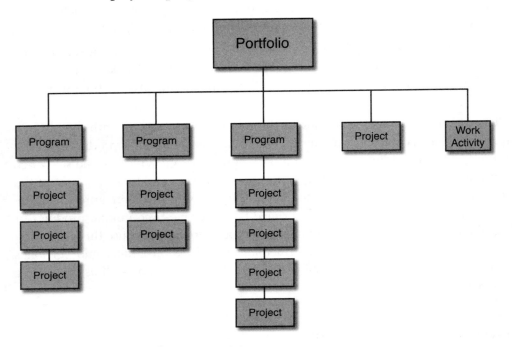

Figure 1.1 Projects, programs, and portfolios

Table 1.1 Project, Programs, and Portfolio Management Structures

	Projects, Programs, and Portfolio Management		
	Projects	Programs	Portfolios
Structure	Single project	Multiple related projects	Combination of related and non-related projects, programs, and work activities
Customer	Single customer	Single customer, multiple deliverables	Single or multiple customers, multiple deliverables
Objective	Specific deliverable	Multiple deliverables supporting a single objective	Multiple projects and programs supporting a business unit objective
Management	Project manager, specific project staff	Program manager, oversee project managers, multiple staffs	Portfolio director, oversee multiple Program and project managers

The project manager must now understand *when* and *how* to use projects and programs within a portfolio to accomplish strategic objectives within the organization. To start, we look at the organization itself and what kinds of products or services the organization is designed to accomplish. As we have already discussed, some organizations produce more one-time unique items that we can now call *projects,* whereas other organizations have ongoing production of products that can be called *programs.* The primary delineation here is as follows:

- **Projects** are unique, have a start and stop, and employ a single objective.
- **Programs** are less unique, have a start, but might not necessarily stop unless the organization no longer has a requirement for that product or service.

Projects can also be used in smaller, less formal ways within the organization—for example, the creation of documentation, a new process, or a one-time activity within an organization such as moving a department or building a new facility. The organization may also use projects to create more specialized items it needs such as research and development, new product introduction, or problem-solving exercises.

It is important for the project manager to understand not only the difference between the terms *projects, programs,* and *portfolios,* but when to use these structures and that there are differences in cost and scheduling resources for these types of structures. Because this text does not go into detail regarding how to build these structures and all the specifics associated with project management, it is more specific to estimating costs, scheduling resources, and controlling budgets and schedules for these types of structures. Another primary difference with these types of structures is the high-level approach taken and how to manage these structures within an organization.

Project Management Versus Program and Portfolio Management

Organizations use projects, programs, and portfolios depending on the type of business they conduct and how they feel each of these structures will benefit the organization. Most organizations structure their operations based on the requirements of managing activities to achieve their objectives. Although it's common for organizations to be divided into departments based on each department's type of activity, profit centers are most likely to use projects, programs, and portfolios to manage activities that produce products and conduct services. Following are definitions of how to manage these structures:

Project management—The utilization of skills, knowledge, and experience to effectively manage resources to complete work activities required to accomplish a project objective.

Program management—The utilization of skills, knowledge, and experience to effectively manage project managers, staff, and resources required to complete multiple project objectives within a defined program.

Portfolio management—The utilization of skills, knowledge, and managerial experience to effectively oversee the completion of program objectives, projects, and work activities required in the overall portfolio objective.

Connection to Organizational Needs

Organizations use *projects* more independently because they are designed for a unique purpose, have a start and stop, and can be viewed within these organizations as a disposable function. This is not to demean the function, purpose, or importance of a project, because projects can be very long in duration, be very complex in structure, and produce some of our world's most fascinating and incredible things; it is the fact that they are used one time for one purpose. After a project is completed, it is rare that the organization returns to a project and continues performing activities.

Organizations use *programs* and *portfolios* on an ongoing basis because they are groups of project and organizational activities that serve a purpose for the program or the portfolio. The primary difference between the program and the portfolio is that the program has a single purpose, and therefore, projects and activities are typically co-related for that objective. Portfolios are typically at a higher level than programs because portfolios can have a combination of projects, programs, and organizational activities that may or may not be related for the purpose of that portfolio.

Project Management

Managing a project requires the manager to structure the project using specific resources for specific tasks and to estimate and control costs for specific tasks to ensure a project is completed on schedule and within the target budget. The project therefore has a very detailed nature, and management of each individual task is very important. Although the project manager may or may not be actively involved in structuring a project or gathering cost estimates, the project manager is responsible for the following tasks:

- Overseeing the completion of each task
- Assessing potential risks and engaging in control activities that keep costs and expenditures on budget and completion on schedule
- Managing resource schedules to complete specified activities

Most organizations appreciate managers having an understanding of the "big picture," but the project manager is typically focused on the individual project and the success of completing tasks to accomplish the objective of the project. Figure 1.2 illustrates a basic project structure of specific work activities to manage in completing an objective.

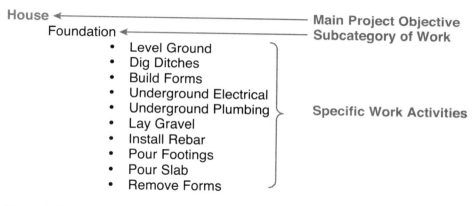

Figure 1.2 Specific work activities to manage on a project

Program Management

Programs have a slightly less specific nature, and organizations view them as an ongoing umbrella of projects and activities directed to a more general objective. Programs can be a group of projects that are all targeted to be delivered to one particular customer (related by customer), or they can be a grouping of projects that are similar in nature, such as a particular product that might be sold to different customers (related by product or service type). Because programs are typically ongoing, projects and activities within the program are added and removed as they are completed, and new related projects and activities are added to maintain the objective of the program. Because managers are assigned to each individual project within a program, the management of the program has a less specific application and is more generalized to the overall success of the program itself. The program manager generally is responsible for

- Selecting projects and activities to support the program objective
- Managing project managers
- Overseeing scheduling of resources and cash flow required for all projects and activities within the program
- Ensuring ongoing commitments of the program are met

Figure 1.3 illustrates how a program is structured using related projects and activities.

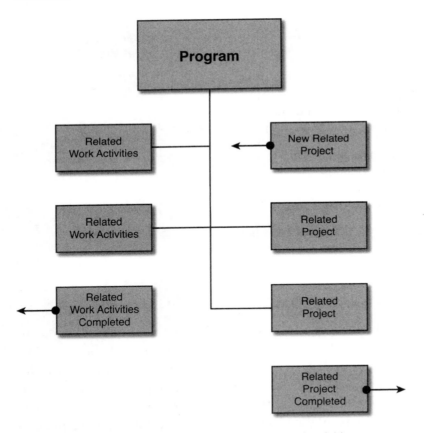

Figure 1.3 Ongoing program with related projects and activities

It also is important for program managers to be involved in scheduling when projects and activities start and stop within a program because this task is important internally for resource and cash flow management as well as externally for managing customer expectations. These mid-level managers typically will find it important not only to understand but also to be actively involved in the "big picture" with regards to their program. This might include direct customer involvement and accountability to senior management within the organization.

Portfolio Management

Portfolio management is similar to program management except that it is at a higher level because portfolios can include programs, individual projects, and activities performed within the organization. Some organizations identify portfolios as departments or divisions within the organization, whereas other organizations might see portfolios as simply higher-level forms of program management. The difference between portfolio management and program management is that portfolios can have programs and projects that are not related and therefore may not share common resources or customer requirements. Portfolio management requires

- Evaluating what programs and projects should fall within the portfolio
- Controlling when projects and activities start and stop within the portfolio
- Determining critical resource and cash flow requirements for programs, projects, and activities to support the overall objective of the portfolio
- Ensuring the ongoing commitments of the portfolio are met

Depending on the size of the organization, the portfolio manager is typically seen as either mid-level, upper-level, or executive management. Figure 1.4 illustrates how several nonrelated projects, programs, and activities would look within a portfolio structure.

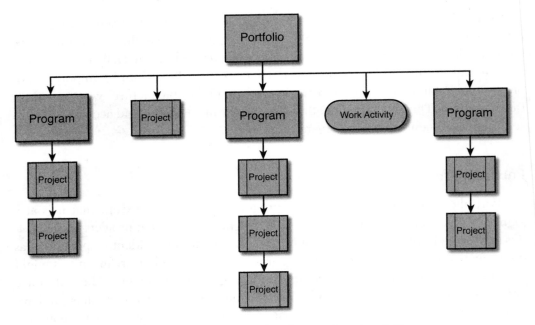

Figure 1.4 Portfolio with related and nonrelated projects and work activities

It is critical for managers assigned at each of these three levels to take these positions seriously because this is how organizations benefit from these types of structures. As you have seen, project managers have a much more specific scope of involvement focused on an individual project; this approach is important because it helps ensure not only that the manager is focused on the completion of the project, but also that the objective is accomplished on schedule and at the budgeted cost. It would be difficult for program and portfolio managers overseeing several projects and activities to be focused on any specific project task or objective, and this is why the function of the project manager is important within the organization.

Likewise, it is important for the organization to group related projects and activities under a program in which the program manager can oversee all activities related to the program objective. At this level, the program manager is focused simply on the objective of that individual program and not at a specific project-level objective or at a much higher division, department, or portfolio-level objective. Both the program manager and portfolio manager have responsibility for resources and cash flow required for several projects and activities within their respective program or portfolio, but also are focused on ensuring the activities within their program or portfolio, on an ongoing basis, are meeting their objectives. Successful organizations use these management styles to bring structure and organization, thus promoting focus, clarity, and success in completing organizational objectives.

Project Life Cycle

Although projects are a compilation of work activities performed in a sequence that, when completed, produce the project objective, projects also go through transitional states called *phases* that characterize the general nature of work being carried out during each phase. These phases are not to be confused with the *project management process groups*, which consist of activities that can be performed throughout the entire project, but are more high-level designations of what is being accomplished in a particular time span. There are four primary project life-cycle phases:

- **Conceptual**—The conceptual phase marks the initial efforts taken at the beginning of a project. In this phase, activities include review of the project objective, customer specifications, and determination of initial stakeholders. Also included are general estimates for resources and facilities; general financial and schedule requirements; and feasibility, risk, and strategic benefit studies. The *project charter* is typically the output of this phase and, if approved, marks the official beginning of the project.

- **Planning**—Now that the initial idea of the project or charter has been approved, more specific details of work activities can be identified and sequentially categorized into work packages and documented in a work breakdown structure. Resource, cost, and schedule estimates can now be gathered and potential risks documented. The output of this phase results in the development of the *project management plan*.

- **Execution**—In the execution phase, the project team perform and complete work activities. During this phase, it's typical to see the highest amounts of resources used as well as financial and facility requirements implemented. Because risk events are typically more associated with work activities, more risk events are likely to occur during this phase. The output of this phase is the completion of the project objectives.

- **Closure**—Closure of the project occurs when acceptance of project deliverables is confirmed. This phase includes reduction of workforce, resources, and facilities as well as completion and sign-off of any contracts and completion of all financial payables and receivables. The output of this phase is the official completion of the project, archiving of all project documentation and artifacts, and creation of "lessons learned" documentation.

All projects pass through these phases, but the purpose of this text is to analyze what effects and influences exist on cost and schedule throughout the project life cycle. Figure 1.5 illustrates these phases and the relative level of work activity and resource requirements encountered within a project life cycle.

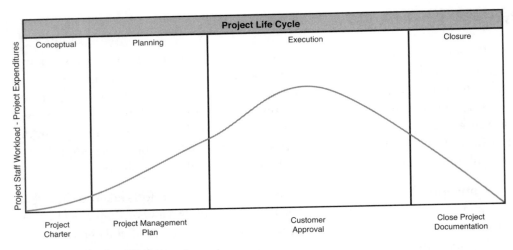

Figure 1.5 Basic project life cycle

Within each phase of the life cycle, the project team has different members, including stakeholders who can have varying levels and types of participation with project activities and decisions. Some phases involve more development and decision making, whereas other phases may be simply conducting and completing predetermined work activities. Each phase of the life cycle presents challenges to the project manager in estimating, monitoring, and controlling the budget and scheduling activities. Figure 1.6 contrasts the different challenges presented to the project manager with regard to budget and schedule through each phase of the project life cycle.

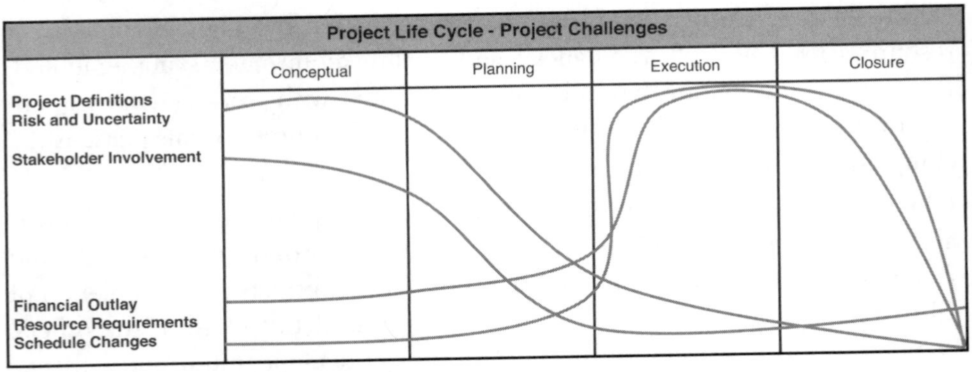

Figure 1.6 Comparison of cost and schedule challenges through project life cycle

Each part of the project life cycle presents different challenges for the project manager to address. In general, the beginning parts of the project (conceptual and planning) hold more challenges for project development, stakeholder management, and risk or uncertainty. In the execution section, all the work activities are managed, and schedule issues and budget management are the most difficult. This is also the point at which change management is at its highest potential and can cause the most conflict and uncertainty if not managed well. The closure section is usually a stressful time because the procurements and contracts are being evaluated for completion and compliance to ensure payments. This is why organizations hire professional project managers who are skilled and experienced in dealing with all these phases of the project life cycle; several areas can have different types of problems, and the project manager will know how to manage these areas.

Review Questions

1. Discuss what is meant by producing and supporting functions within an organization.
2. Explain the connections that projects, programs, and portfolios have with organizational needs.
3. Explain the differences between projects, programs, and portfolios.
4. Discuss the main areas in a project life cycle.

2

Initiating Process

Introduction

In the first part of the project life cycle, during the conceptual phase, the initial activities required to create a project are conducted; this step is called the *initiating process*. This process is considered to be one of the most critical elements of a project because many aspects of the project are defined at this point. Although several areas of the project are addressed during this process, the primary goal of the initiating process is to capture enough information about a potential unique endeavor to allow key individuals to decide whether to proceed with project activities.

Whether projects are small and simple or large and complex, detailed information is gathered on specific activities required within a project. To decide whether a project is feasible and should continue at the beginning, key individuals should complete the following seven general elements during the initiating process:

- Identify a need, problem, or opportunity
- Define the project scope
- Identify a project objective and deliverable
- Determine general estimates of cost, resources, and time required to complete the objective
- Identify internal and external stakeholders
- Assign a project manager
- Commit initial financial resources required to complete the objective

Information gathered during this process is more generalized, provides less detail, and is initially compiled for high-level decision making only. We address six areas within the initiating process that have the greatest influence to cost and schedule estimating as well as project control. Figure 2.1 shows the initiating process and corresponding outputs during the conceptual phase of the project life cycle.

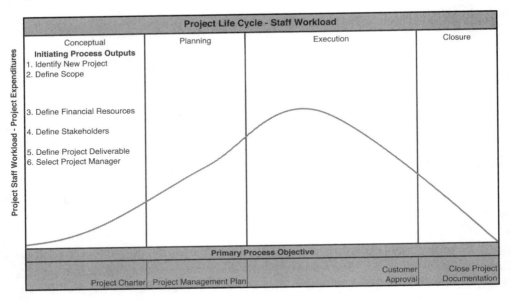

Figure 2.1 Initiating process

Project Origination

The basic business model of supply and demand based on a supplier/customer relationship also forms the basic model of a project. Because projects are a unique endeavor to produce an output deliverable and have a start and completion point, the output deliverable is defined by a customer need and will be produced by the project team acting as the supplier. This supplier/customer relationship can be formed between two parties within an organization or can be established externally between two organizations. Thus, the origin of a project begins with a need for a unique deliverable or service defined by a customer and recognized by a supplier as an opportunity. Individuals representing the supplier then further define the objective and perform an analysis as to the feasibility of the opportunity. If the supplier deems the opportunity is feasible, the supplier notifies the customer that it will pursue the opportunity and, in doing so, begin creating a project.

Internal Projects and Improvements

Within some organizations, internal departments have the resources to produce unique deliverables that can be used or will be required by other departments within the organization; these are called *internal projects*. A distinction must be made in that some internal departments are created to support the operations, and their normal daily tasks are not necessarily unique. One example is the IT department within an organization that is tasked with maintaining network servers and setting up office cubicles for employees. Because these tasks are done on a regular basis, they are not unique and therefore not considered projects. If the IT department were to embark on installing new servers that required hardware installation and testing to verify they were working properly, this task might be considered a project because it is unique to the department's normal daily work activities.

Departments within an organization may determine that they require something new developed or an improvement that is outside their normal repetitive operation; these can be considered projects. In this case, a team of individuals within the department performing the activities of the project would be considered the project team, and the department in its entirety making use of the project deliverable would be considered the customer. If two departments are in collaboration, the department providing the solution or deliverable would be considered the project team or the supplier, and the department receiving the deliverable would be considered the customer. It is common for larger organizations to utilize internal resources to supply unique deliverables for other departments within the organization, therefore creating internal projects.

There are advantages and disadvantages with organizations using internal resources for internal projects. Management has to analyze the cost benefit to determine whether this is the best course of action versus hiring an outside contractor. A major advantage for organizations using internal resources is that contracts do not have to be drawn with external suppliers, and managers in charge of the project might have more flexibility with functional managers in the use of resources to complete projects. Another major advantage is the availability of human resources, equipment, and materials that are not being used at 100 percent capacity; they can be used for project activities, thus making it less expensive than hiring external resources.

One disadvantage with the use of internal resources is that resources may not be capable of providing either adequate skill or quality with individual activity assignments. The result is that the output deliverable does not fully meet expectations of the receiving department. Another disadvantage is that the use of internal resources for project activities requires they are taken off their normal activities and the resulting impact this can have within their department in normal daily operations.

It is usually best for the organization to first consider the use of internal resources to complete project activities for efficiency and lowest cost before soliciting external resources to complete project activities. The warning here is that although human resources may be available and appear skilled, there is the risk these resources will not be as efficient in completing project activities. This results in a substandard deliverable, which costs the project more in requiring extra time or resources to complete project deliverables to meet customer standards.

External RFPs and RFQs

In response to market demands, many organizations provide unique products or services that other organizations could benefit from. Conducting business in the marketplace is about establishing relationships between organizations. There will always be an organization that has something to offer while other organizations require products or services, thus forming the business-to-business relationship. This connection can be established through an organization preparing a proposal suggesting how it would meet a specific customer demand based on an advertised requirement from that customer; this is called a *request for proposal (RFP)*. The organization broadcasting the requirement is in complete control of this scenario because it is soliciting and reviewing proposals and will select the supplier it feels is best qualified to complete the objective.

After a proposal is accepted and a connection is established between two organizations, information from the proposal is used as a foundation for defining a new project. The proposal is a very important document because it explains how, what, and why an organization plans to complete the objective in general terms. Although the proposal has a great deal of information about the organization and its capabilities, as well as examples of previous products or services, it generally includes little detail about the specific objective itself. There is only enough information about the objective to ensure the customer is confident the organization understands what is required, and this is the point at which good lines of communication need to be established.

The proposal, then, is used more as a selling feature for the organization, and not as a way to document all the details of the objective. If the customer feels confident the organization can complete the objective, this opens the door for more communication about the specific details of the objective. Consequently, the proposal served its purpose as a tool to win business. It is now incumbent on the initial stakeholders, which may include the project manager, to begin gathering specific information on the objective that will be used to form the beginning of the project.

In other cases, organizations that provide a product or service might be approached by a customer in need of their product or service, but they require modifications to a standard product that will make it unique and will be managed within a project. The customer in this scenario would ask a specific organization to give price and availability of the product or service, including the modifications; this is called a *request for quote (RFQ)*. In this case, the organization providing the product or service is in complete control of this scenario because the customer is inquiring about a specific product or service and the availability of modifications that may be possible. In most cases, the request for quote includes more detail about specific products or services and any special conditions that will be required. If an agreement is made, this document can serve as a more detailed foundation in information gathering at the beginning of the project.

Project Stakeholders

Organizations that are either heavily projectized or more traditional, having a functional structure, do not manage projects in a vacuum, having no influences of any kind. Projects always have some form of connection to personnel that represent management, procurement, development, or customers and possibly other departments that can and will have an influence on a project. Individuals who can directly influence the development, management, or final outcome of a project can be considered stakeholders. *Stakeholders* are formally defined as any individuals, whether internal or external to the organization, who have a stake in the project.

Project Stakeholder Management

Depending on how projects are started, some stakeholders are selected at the beginning of the project before the project manager is selected. In other cases, the project manager is included in the project development from the beginning. At what point the project manager is selected on a new project is critical because this plays an important role in not only the development of the project, but also in the project manager's understanding of how to manage the stakeholders and their expectations. There are three elements regarding stakeholder involvement that the project manager will be responsible for:

Identify—The first part of stakeholder management is identifying who are stakeholders. These individuals typically include those who "have a stake" in the project. They can be involved in the initial development of the project in negotiations, specification development, and approval of the charter. They also include those with decision-making authority.

Manage stakeholders—The second element is to define how stakeholders will be managed. This step does not include having responsibility for authority over them, just how they are involved with the project and the general communication plan. Because these individuals have some level of interest in the project, the project manager needs to determine how to manage their expectations of the project.

Manage stakeholder participation—The third element is to identify what the stakeholders will be doing in their involvement and how to address issues as they arise.

The project manager has ultimate responsibility to the organization as well as the project to understand who stakeholders are, why certain stakeholders are associated with the project, how to manage their needs and expectations, and how to develop appropriate and effective communication channels. If stakeholders ultimately have a stake in the project, they will also have influence on some part of the project or require information or deliverables from the project. Figure 2.2 shows stakeholder relationship to a project.

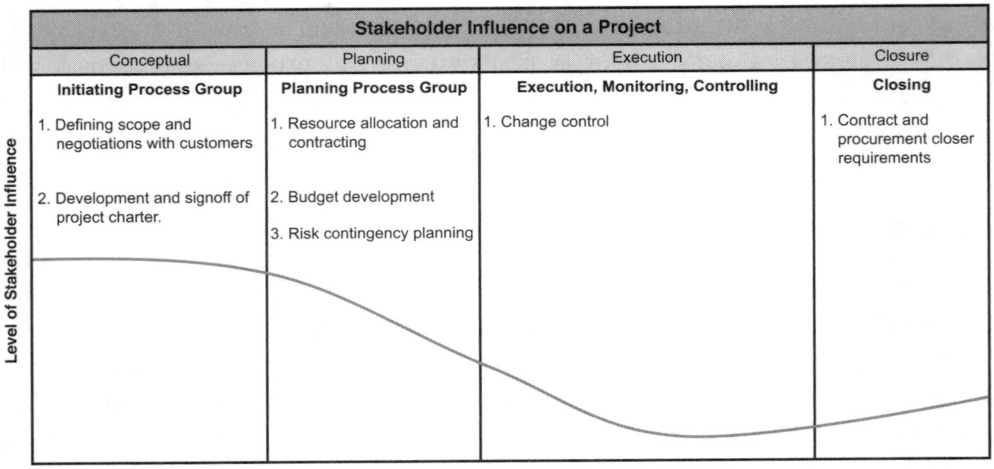

Stakeholder Influence on a Project			
Conceptual	Planning	Execution	Closure
Initiating Process Group	**Planning Process Group**	**Execution, Monitoring, Controlling**	**Closing**
1. Defining scope and negotiations with customers 2. Development and signoff of project charter.	1. Resource allocation and contracting 2. Budget development 3. Risk contingency planning	1. Change control	1. Contract and procurement closer requirements

Figure 2.2 Stakeholder-to-project relationships

Project managers need to understand the details of how stakeholders can influence the project. This is done with a stakeholder analysis. Information to perform a stakeholder analysis starts with the stakeholder registry, which lists the stakeholders, their responsibilities, and what type of influence they might have on a project. Project managers gather other forms of information to determine whether stakeholders are receiving information from the project or giving information to the project. They also determine the expectations of the stakeholders in what type of communication will be most effective in communicating information to and from them. The stakeholder analysis is the primary tool project managers use to manage stakeholder influence on the project and therefore should be performed as early in the project as possible.

Expectations

One of the most difficult areas project managers have to manage is the expectations of stakeholders. As you have seen, expectations can be the influence stakeholders have on the project where information is coming to the project or the information or deliverables produced by the project are going out to the stakeholders. If a project manager has ultimate responsibility of the project, he must also understand the relationship stakeholders have to the project and the impact that influence can have on the project. In many cases, the project manager must walk a fine line between accepting influence as a result of a stakeholder's managerial level versus rejecting influence that is unwarranted and may negatively impact the project or conflict with project objectives. The project manager must understand who each stakeholder is and what responsibility that stakeholder has in regard to the project.

Likewise, stakeholders such as upper management or customers may have expectations in regard to project progress, status reports, and any critical information regarding the deliverable that is being produced by the project and being reported out. The project manager must design communication strategies for what, when, how, and why certain information is communicated to these types of stakeholders. The project manager must be careful in managing stakeholder expectations in delivering only the appropriate information in the correct format as required by the stakeholders to fulfill their expectations.

In either case, the project manager must know how to identify stakeholders, develop a plan to manage the information either coming from or going to the stakeholders, and understand what impact the stakeholders can have on a project's cost and schedule. The focus here is to help the project manager understand the impact stakeholders can have on both the budget and scheduling of project activities. It is also critical for the project manager to understand the importance of developing stakeholder management plans and how this can improve the relationships between the project manager, stakeholders, and project staff.

Project Selection

Organizations will find that decisions must be made every day concerning projects that may be product related, may be process development related, or may improve or expand the organization. Managers at all levels are evaluating what projects will benefit the organization and when and how to start projects so as not to disrupt day-to-day operations. This can be difficult because some projects may be very important and require several people to analyze and improve, whereas other projects may be small and managers can easily start and finish them to accomplish something important for their department. In any case, projects have to be evaluated based on information that was gathered, and selection criteria have to be developed for managers to properly assess what projects will benefit the organization most and the best timing to start projects.

Although it might be exciting for managers to assess which projects might be beneficial for the organization, they usually have to consider constraints that may or may not allow certain projects to commence. Because most project ideas were created from a need within the organization, not all projects are technically feasible, are financially equitable, or have a return on investment that justifies the project objective. To help managers decide what projects are best for the organization, they must gather information and develop selection criteria for projects to be analyzed and approved. The following sections explore the constraints that organizations may have in being able to carry out work activities for a project and present selection methodologies that give managers tools to evaluate which projects organizations should approve for maximum benefit in return on investment. Most organizations have constraints at two levels: the organizational level and the project management level.

Organizational Constraints

Depending on how they are structured, organizations can present constraints to project selection as well as promote best practices in project selection. Organizations may also have many resources that can be utilized in project selection, but these same resources can also present constraints, making the selection process less efficient. Six primary areas within organizations can present problems in project selection:

In-house technology—As organizations evaluate what types of projects will be started, one constraint can be whether or not the organization has adequate technology to produce the project deliverables. Some projects, either product or process improvement, may fall within the organization's technological capability and will not pose any constraints. Other projects or specialized product development may require technology the organization does not have, and this may require the organization to acquire this knowledge or simply decline the project due to a lack of technology. It should be noted that when organizations require new technology, this is how they improve and expand their capability, so this requirement may not always be seen as a constraint, but an opportunity. In other cases, when organizations select a project that requires technology they do not have, they may find it difficult to produce project deliverables and put the organization at risk of not completing the project objective.

Human resources—Project work activities, in most cases, require certain skill sets of human resources. Whether or not the skill sets are available within the organization can impose constraints on completing project work activities. It is incumbent on those reviewing the information within the statement of work on a potential project to have knowledge of the human resources available so that they can make accurate assessments of these resources to complete work activities. It is unfortunate when organizations do have the skill sets available, but those in the selection process are unaware and do not take a project as a result. In other cases, individuals in the selection process are aware of unique skill sets within the organization that others may not know of, and the unique opportunity is realized based on the knowledge of a certain skill set. Consequently, a critical element for those in the selection process is to know the skill sets available within the organization because this can present constraints as well as opportunities. In some cases, an organization may want to pursue a project that requires skill sets not available within the organization, but it may be willing to hire consultants for the purpose of completing specialized work activities so that the organization can take advantage of the opportunity to complete the project objective.

Management—In most cases, organizations utilize some level of upper management in the project selection process, and this is another area that can represent opportunity or constraint. It is critical that those identified for the project selection process are skilled at project selection, are knowledgeable of the resources available, are knowledgeable of the work activities that will be carried out, and understand the overall strategic objective of the organization. Management must also have team skills that allow them to work cooperatively and efficiently with other managers or project selection staff. It is important that managers understand the importance of project selection and the correct evaluation process is utilized so as not to reflect an individual manager's agenda over the goals of the organization. In many cases, it is best to use management in project selection, as they have the unique perspective of understanding the bigger picture of organizational goals and the knowledge of skill sets or abilities of staff for potential project teams. Other constraints may be seen in upper management in that they might not see eye to eye on the selection of certain projects. It is important for upper management to understand their influence in project selection and that they acknowledge the importance of the effective selection process.

Facilities and equipment—Another important area to factor into the project selection process is the general infrastructure of the organization and its capabilities for handling projects outside normal daily operations. Most organizations have the facilities, equipment, materials, and human resource staff required to carry out the daily operations of the organization but may or may not have extra capabilities to carry out projects. It is important for those in the selection process to determine what infrastructure resources will be required in addition to daily operations to manage project work activities. It is important to understand the difference between having extra resources in addition to daily operations for project activities and borrowing resources from daily operations for project activities. Those carrying out the selection evaluation should communicate with functional managers in determining what resources are available. It may be determined that resources are not being fully utilized and therefore can be used for special project activities. It may also be determined that resources required for a project are being used at full capacity for daily operations and extra resources will need to be brought in to carry out project activities. This is an important factor in the selection process and not only the availability of facilities equipment and human resources, but the cost of bringing in extra resources if needed.

Financial resources—Depending on the size and complexity of a proposed project, part of the selection criteria needs to be an assessment of how much financial backing the organization will need to produce to carry out project work activities. If an organization is evaluating a project such as the creation of new documentation or a process improvement upgrade, this may require little, if any, financial resources and simply require human resource labor time. Other large-scale and more complex projects may require a great deal of infrastructure to be purchased and/or rented, materials to be purchased, and human resources to be brought in to complete work activities. These types of projects can be very expensive and may require specialized funding. This will typically be noted during the selection criteria as to the ability to adequately cover the financial needs of the project. In the selection process, it is important that financial commitments are made before a project has begun so that cash flow will not be an issue for the organization and a lack of finances will not become a constraint during the course of the project.

Functional, matrix, or projectized—How organizations are structured can play an important role in how projects are selected, implemented, and managed. Functional organizations have a more traditional structure and are divided into functional areas or departments such as administration, engineering, procurement, manufacturing, and inventory control. Projectized organizations have an organizational structure design around major projects in which the organization is engaged. Representatives of many traditional areas within the organization are assigned to project teams, and after they complete their work activities on one project, they are reassigned to another project. The organization may appear to have the traditional departments, but most of the employees in the organization perform their work assigned to projects and move from project to project. Matrix organizations have a blend of both functional and projectized type structures where the organization may have traditional departments and employees reporting to functional managers, projects are still a large part of the organization's business, and employees spend portions of their time on projects and other parts of their time performing work activities for the functional department.

Functional, matrix, and projectized structures each have their advantages and disadvantages in regard to project selection. Functional organizations may find it more difficult because projects are not a fundamental component of the business and therefore require special effort to perform project selection processes. Projectized organizations typically have a well-refined and efficient project selection process because projects are routinely evaluated as the major component of the organization's business. Matrix organizations have a blend of both functional and projectized structures where they have traditional departments, and staff may report to functional managers or project managers, but the organization has performed many projects and therefore is comfortable with project management and project selection.

Typically, projectized organizations have a more streamlined and well-developed project selection process because this is their primary business; they also typically have staff employed for the purpose of information gathering, project evaluation, and selection. Functional organizations, although they may occasionally conduct a project, are typically seen as being less efficient at project selection. This is due to the focus being on their primary daily business model, which rarely includes special projects. This results in a lack of qualified staff involved in accurate information gathering and the development of an effective project selection process. Matrix organizations fall somewhere in between functional and projectized structures depending on how many projects they engage in and how well developed their selection processes have become based on those projects.

Project Management Constraints

The next primary constraint is how well an organization is prepared for projects. Managing projects can present great opportunities for organizations or can be their worst nightmare, depending on how the organization is structured. Critical areas for project success include staff skilled at information gathering, project selection, and project management, and staff employed to develop a project management office and effectively manage portfolios and programs. Larger functional organizations may include the staff due to the size and complexity of projects conducted in their organization. Organizations that do not typically perform projects may struggle if lacking these types of specialized project management resources. Some organizations may find it more effective to educate and train functional managers in project management skills and processes to more effectively manage projects. In any case, having staff skilled in project management techniques and processes helps the organization more effectively select and manage projects, allowing it to capitalize on unique opportunities. Most organizations find that constraints at the project management level fall into the following two general categories:

> **Project management maturity of the organization**—As you have seen, organizations can vary in the amount of experience they have with conducting projects, and this can play an important role in project selection and management. One of the more fundamental keys to successful project selection is how much experience the organization has in selecting and conducting projects. Individuals who have been given the responsibility of gathering data and analyzing potential projects for selection typically draw on past experience and the history of projects that the organization has completed to find pieces of information that may influence the selection process.
>
> Individuals who have participated in prior projects within the organization can also provide critical information during the selection process. Information can be in the form of technology required during certain work activities, information about customers, issues from previous projects, and lessons learned that can be used in the analysis for project selection.

Another critical component that defines the maturity of an organization, relative to projects, is the availability of skilled project management staff. Project managers play a vital role in the success of selection and management of projects because they are trained and skilled in project management techniques and processes, thus allowing them to be effective at efficiently completing projects. If an organization lacks skilled project management staff, this can be a serious constraint for effectively managing projects.

If an organization has a well-developed project management process, it typically has key components such as a project management office (**PMO**) or established project management templates and processes used to standardize project management within the organization. If several projects are running simultaneously, some organizations employ managerial staff skilled in developing and managing programs and even larger multiprogram portfolios. These organizations are considered to have a higher level of maturity in project management development because these types of structures and processes promote a more efficient project selection and management capability.

Number of projects in cue—Most organizations—whether small with less project management development experience or very large with a more sophisticated project management structure—find that having a number of projects running at any given time can present challenges and constraints. When organizations consider projects for selection, they must have selection criteria that include the organization's capacity limitations and ability to add more projects. This can be in the form of human resources, capital equipment, and materials available, but also must take into consideration skilled project management staff to document and develop as well as run multiple projects effectively and efficiently.

With smaller organizations that have a less-developed project management structure, this lack in structure can be a constraint and is especially important if there is a lack of skilled project managers. These types of organizations may have a great opportunity and all the resources in place to conduct a project but lack the project management capability to properly document and effectively manage a project through to completion. Smaller organizations can also have resource limitation problems such as the availability of human resources, capital equipment, facilities, and financial backing for projects. In some cases, during the selection process, smaller organizations are overly critical of projects that can impose a strain on management in having to make a final decision on which project to select. This type of constraint can also be compounded by the lack of project management experience during the selection process.

Larger organizations that have a well-developed and sophisticated project management structure in place also have to pay close attention to the selection of projects and how they are added into the priority of other projects. Specialized staff called *portfolio managers* or *program managers* are employed to consolidate multiple projects into like programs or portfolios and manage them in groups of projects. It is their responsibility to assist in project selection, placement of projects within programs or portfolios, and prioritization of projects given the availability of resources. These managers typically decide how big portfolios or programs can be given the organization's structure, resource availability, and alignment to strategic objectives of the organization. Figure 2.3 illustrates a structure using portfolio and program management.

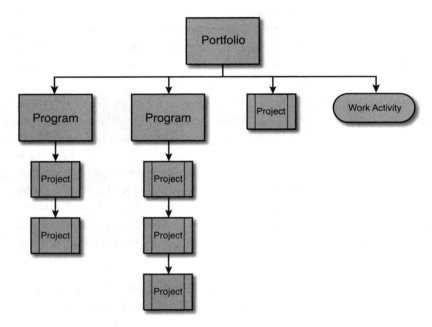

Figure 2.3 Portfolio and program management

Project Selection in Organizational Strategy

If organizations choose to conduct projects, it is critical that they establish a selection model to determine what projects are best suited to meet organizational objectives. Organizations typically have questions as to why a project is selected and the methodology as to how projects are being analyzed, but they must establish a selection model that considers these general factors: the *type of organization* and how many projects are being evaluated; and the *type of data being analyzed* and whether projects will be *assigned to programs or portfolios* or be assessed as an *independent project*. As you have seen, there are two primary reasons projects are selected based on the *type of organization:*

Project comparison to other projects—When organizations use projects on a regular basis, they have typically predetermined that projects are a fit for the organization and spend most of the evaluation comparing projects with each other. This requires information gathering for several projects, and in some cases, they select multiple projects but evaluate the priority and the placement of those projects within the organization.

Process or product improvements for the organization—When organizations do not typically use projects very often, most of the evaluation is to determine the value the project will bring to the organization and whether the organization has the capability to perform project activities. In some cases, these organizations use projects simply for internal process improvement or special unique product development and do not have much overall impact to the operation from a resource standpoint.

In other cases, organizations may use projects to manage large or unique items such as moving the organization to another location or adding a brand-new facility or some form of large-scale activity the organization wants to complete. In these cases, the decision is to use internal resources or contract external professional project management resources to conduct these types of projects. These types of organizations typically find themselves lacking in project selection and management skills, making it more difficult to complete the selection process. It is also common for organizations to assign a functional manager to oversee internal project activities.

After an organization determines why it is selecting a project, it must establish a methodology using predetermined criteria for which to grade projects. Each organization has to evaluate its own business model to determine selection criteria and the prioritization or importance of certain criteria. The *type of data being analyzed* determines the methodology of the analysis used, which generally falls into one of the following two categories:

Qualitative—Qualitative grading is used when little or no actual numerical values have been gathered on project details and decisions have to be made based on more subjective general descriptions. This can be expressed in forms such as high, medium, or low; hot or cold; large or small; or other descriptive type assessments. This type of selection can also include elements of the business such as projects that naturally fall in a program, customer-expected projects that just happen regardless, and other forms of project selection that do not have numerical analysis. Although this form of grading can be effective, it should be used only when actual numerical values are inconclusive or not available at all.

Quantitative—Quantitative project selection utilizes more specific numerical data gathered on the project activities. Actual numerical data such as size, quantities, temperatures, and financial data can be used more accurately to objectively assess projects for selection. It is highly advisable that individuals use a quantitative analysis when possible because it can produce a more accurate assessment of project details as well as establish starting points for the development of the statement of work.

Organizations that use projects on a regular basis, in most cases, have several projects in process simultaneously. These types of projects are a function of the organization's normal daily operation and reflect the organization's strategic goals. These projects are usually the result of specific customer requirements for unique products or services. Organizations structured to respond to multiple customer requirements consolidate projects into groups called *programs* or *portfolios*. Other organizations that are not structured for multiple projects based on customer demands, but manage projects independently of each other, assess projects individually based on their value to the organization.

Portfolios and Programs

Organizations structured to manage multiple projects divide these projects into groups called programs and portfolios. As covered in Chapter 1, "Basic Project Structure," programs are the grouping of like projects so as to make the most efficient use of resources within an organization. Portfolios are typically used at a higher level to group multiple programs and can include individual projects and single activities that might be unrelated but are managed within the portfolio. How the organization divides projects into programs or portfolios is important in the project selection process because criteria for evaluation include how the organization categorizes and prioritizes projects. Organizations using programs and portfolios find it more effective to group projects into the following types of categories:

Customer based—If an organization produces a more specific type product and customers require several unique variations of the product based on their application, the organization groups these unique product projects into a program for specific customers. This approach is more typical of companies that may have a small number of customers and customers who require very specific and unique variations of the company's product. For example, a company that produces cell phones groups the production of unique variations of cell phones based on the specific needs of each customer, such as AT&T, Verizon, and T-Mobile. This way, organization can manage all the projects for that specific customer and assign a program manager as the point of contact to manage customer expectations. This approach also allows the program manager to manage internal resources based on the prioritization of projects to meet customer needs. Usually, customers prefer to have a single point of contact who has knowledge of all their individual projects because this promotes better communication and consistency of project details and status reports. Project selection criteria in this environment have more focus on customer needs, expectations, and prioritization within each customer's program. Figure 2.4 illustrates the structure of an organization based on customer programs.

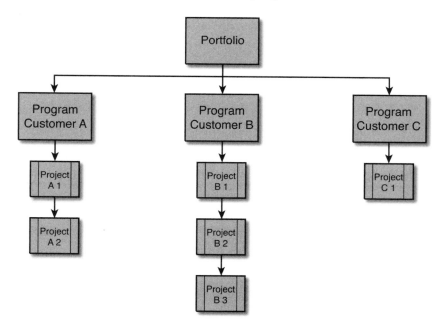

Figure 2.4 Customer-based programs and projects

Product based—Other organizations may choose to group projects by product or service type where the organization may offer a small number of products or services and have a larger customer base. In this case, the organization looks to streamline operations by consolidating resources required for each type of product. For example, a company that produces farm equipment might group like products into programs, such as all the farm tractors in one program, farm tractor accessories into another program, and tree-harvesting machines and ground-harvesting machines into separate programs. For each program, the company has unique variations of each piece of equipment created for various customers, but the organization can consolidate specialized resources and equipment required for each type of machine to streamline the operations. Selection criteria for this environment are focused on the product and the overall value a product change or addition will bring to the overall business. Figure 2.5 illustrates the structure of an organization based on product programs.

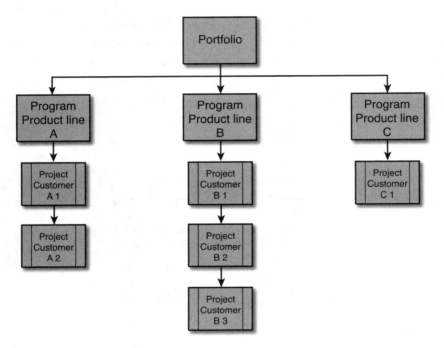

Figure 2.5 Product-based programs and projects

Organizational division—Larger organizations that do business in multiple unrelated markets divide operations into separate business units. Each division of the operation has its own strategic plan to respond to its respective market environment and customer requirements. For example, Siemens has four primary divisions, including Energy, Healthcare, Industry, and Infrastructure and Cities. Each of these four major divisions or portfolios of business has several programs, projects, and individual work activities that are specific to that portfolio and are not used within other portfolios. This allows the organization to acquire specific resources, equipment, and facilities unique to each portfolio of business, making the overall business more efficient at responding to market demands. This also helps portfolio managers focus on project selection for specific programs within their portfolio. Figure 2.6 illustrates the structure of an organization based on organizational division.

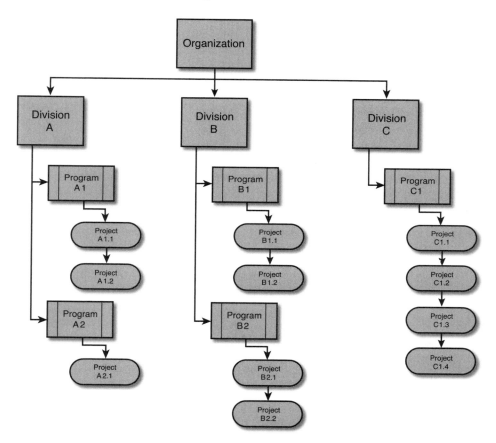

Figure 2.6 Organizational division–based portfolios and programs

When managers are developing selection criteria for projects assigned to programs or portfolios, other factors such as prioritization, customer relationship, and expectations become important criteria to include in the selection process. In many cases, having well-defined programs and portfolios can actually make project selection easier because they are more focused to either customer-, product-, or portfolio-based business.

If the organization does not typically use projects in the course of its daily operation, the selection process assesses each project on an independent basis for the value it brings to the organization. Depending on the size of the organization and the complexity of the project, independent project selection might be easier or much more difficult than selection of projects used in programs or portfolios.

Independent Projects

Projects that are used for specialized activities such as research and development, process improvement, or infrastructure-related business activities are called *independent projects*. In most cases, independent projects are developed out of a unique and specific need within the organization to accomplish something that is not necessarily associated with the daily operation of the business. If an organization chooses to expand its operations, move to a different facility, or install new equipment, it might organize such activities in the form of a project and assign a project manager to oversee all work activities to completion. These types of projects have similar selection criteria as other types of projects but may include, for example, assessment of different buildings and locations if the organization is expanding the operation, decisions on how to expand, or the application and return on investment of a research and development project.

With individual projects, it's also important that selection criteria also include the capabilities of the organization and management of the project and resources available to carry out project work activities. As you have seen with these types of projects, organizations can struggle in properly documenting project objectives and details for project deliverables, as well as the availability of project management staff and resources that do not impact daily operations.

When organizations do not perform projects very often, they lack the skill in effective and efficient project management, and projects suffer as a result. Project selection for these environments should weigh the size and complexity of a project with the capability of the organization to efficiently carry out that level of project. It might be determined that hiring outside professional project management resources is the best course of action if a complex project needs to be completed. If a project assessment reveals the scope and complexity of a project is small enough that an internal functional manager can effectively oversee project activities, this information can play a vital role in the selection process. In many cases, it's common for organizations to underestimate the amount of work required, thus improperly managing projects. If an organization does not have skilled resources in project management, this is the reason projects can have cost overruns, schedule delays, and foreseeable but unplanned risk events that result in poor project performance and unsatisfied stakeholders.

Selection Models and Methodologies

Executives and managers regularly evaluate project proposals for change or the creation of something new that requires information to be gathered, analyzed, and decided upon, and this requires tools to properly analyze project data. Some data is more generalized, using little detail and subjective description that requires certain types of analysis; whereas other data is more specific, numerical, and more objective, requiring tools better suited for that type of data analysis.

Five basic elements of project selection models are common to the selection process, regardless of how much actual analytical data is available:

Realism—The model should take into account the overall strategic alignment of a potential project so as to ensure the benefits of the completed project built on the business strategy. The model should take into consideration general levels of risk, financial exposure, and availability of resources and facilities to complete the potential project objective.

Capability—The selection model should require a review of the organization's capabilities, level of technology, and general fit in the current business portfolios of projects. This requires the evaluation of projects that are "like" the ones currently being carried out but does not discount the opportunity to expand the capabilities and technology of the organization.

Flexibility—The model should be easily adaptable to modification if required for a specific type of evaluation. This can include changes in regulations, rates, or the general level type of information available to use in evaluation.

Ease of use—The model needs to be easy to format with available information, easy for the project manager and others to derive a conclusion, and easy to communicate data and conclusions to others who need to review the selection model. The model and corresponding data might need to be reviewed at several locations and countries and need to be easy to interpret the evaluation process.

Cost—Selection models should also be in a form that is cost effective to acquire and maintain and also promote continued use by the organization. Selection model tools may, in some cases, be included in enterprise software management platforms currently being used by the organization and may simply have to be discovered, trained, and used. The simpler and more cost-effective models can stay, the more they will get used.

Qualitative Screening and Scoring Models

The quickest and simplest project selection models are qualitative in nature; they are called *scoring models*. These types of models allow the evaluation team to be more subjective in "grading" various criteria for potential projects, as shown in the following examples. The scoring model is used when actual numerical data is not available or for quicker and easier evaluation if selection is required before numerical data can be evaluated. Although these techniques are easy and quicker to use, they are not as accurate and should be used only in general potential project evaluations.

> **Scoring model**—This model uses grading criteria to place a subjective value to a list of items of interest selected by the evaluators, as shown in Figure 2.7. This model can also be structured to use numerical values (1–10) and add weights to further refine clarification of grading for specific criteria.

Scoring Selection Model			
Selection Criteria	Project A	Project B	Project C
Strategic Fit	L	H	H
Cost	H	L	M
Technical Feasibility	M	M	H
Profit Potential	H	M	H
Risk	M	L	H
Performance Grading: L = Low M = Medium H = High			

Figure 2.7 Scoring selection model

> **Bubble diagram**—The bubble diagram evaluation is also a non-numerical way to grade or "classify" projects based on criteria but in a different graphical form, as shown in Figure 2.8. Here, three data points can be evaluated, such as project cost, profit, and the size of the bubble as the relative level of risk. This type of evaluation can be performed easily in a program such as Microsoft Excel.

> Other forms of project selection models like the scoring and bubble diagrams include comparison models such as the *Q-Sort comparative model* (Souder, 1984), *pair-wise comparison* (Martino, 1995).

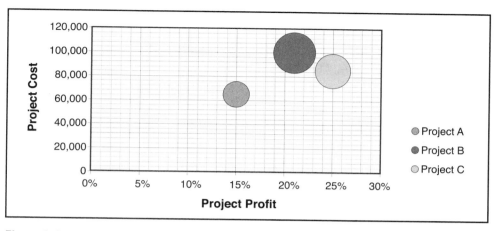

Figure 2.8 Bubble diagram selection model

Quantitative and Financial Models

The second selection method is using a quantitative model that uses numerical data in evaluation and comparison for project selection. In most cases, these models evaluate criteria such as cash flow, cost of the project, rate of return, and the value of money over the life of the project. Because the selection of projects is important, the impact a project will have on the organization needs to be understood so that all who are responsible for the successful operation of a business can be more confident in what the expected outcome of each will be and why projects are selected over others. These types of selection models are more accurate and should be used as the primary tools for project evaluation and selection. Examples of these types of evaluations are as follows:

> **Time value of money**—Depending on the size and complexity of projects, project and financial managers need to be aware of issues that are more elusive, one being the length of a project and the value of money at that point in time. If projects are evaluated in today's value of money, what is that money valued at in one, two, or five years from now when a large project will still be in progress? Does it have the same buying power? Will it yield the same profit dollar value to the organization? These questions need to be answered at the beginning of the project or before the project is selected to better understand the outcome of the project and whether it is actually going to be as successful as it appears in today's value.
>
> Projects have two primary impacts to the organization's financial resources: effects that inflation has on the buying power of money and the inability to invest

monies tied up in the project. Because organizations can invest money in many ways to reap gains, investing in a project ties up money until completion and the profit is realized. If this takes place over the course of several years, what kind of return is that initial investment going to yield, and would that actually be less of a moneymaker than more short-term investments? This is why shorter-term projects are more attractive—the net profit is available quicker and can be reinvested in other projects. Profits are made at a faster rate on several short projects versus a larger profit on a very long-term project. The value of profit is higher on short-term projects because there is a short time component, so the value of profit is not diminished over time. A profit margin can be established on a long-term project, but the value (buying power) of that money is less. The key to this evaluation is in what is called the *payback period*.

Payback period—When projects are evaluated for selection, one primary criterion is how fast the profits can be returned to the organization. Although this question may be simple, it is not answered easily. Profit can be calculated fairly easily, but to get an accurate assessment, one needs to consider other factors such as how long it will take to recoup monies and at what value that will money have in buying power. An example of the effects a project can have on the return on investment given a payback period can be seen in the example of a project with an initial investment of $100,000 and a duration of four years. The payback period is as follows:

Year 1 = $11,000
Year 2 = $19,000
Year 3 = $50,000
Year 4 = $17,000

As you can see, the profit is $17,000 at the end of the project, but it took four years to get that $17,000 value, and it will not have the same return value as it appears in today's value to the organization. Furthermore, if cash flows in years one, two, or three change, the return of the original investment will be extended, further devaluing that money as well as any profits. The basic percent of payback is a rough estimation only for a first-level evaluation that does not include any influences such as inflation. Payback period is a crude way to get a quick answer on the general payback from a rough project selection standpoint and should be used only in that context and not for actual financial projections of a potential project.

Net present value (NPV)—NPV is a more commonly used selection tool because it is an accurate way to evaluate project returns, as it incorporates a discounted cash flow to offset the initial investment. This value helps an organization better understand what the present value of future money looks like to give more refinement in the project selection process. The first step is to calculate the present value using a discounted cash flow, which is simply the expected payoff and a "cost of capital" rate given a time period. You can now say that F is the expected payoff at a time period 1 (you use one year) and r is the cost of capital rate.

The calculation for present value (PV) is as follows:

$$PV = \frac{F}{(1+r)}$$

If you have a project with an initial investment of $100,000 and an expected completion payoff at $125,000 one year from now, you need to know what the present value of that profit is in the project selection. You have other project options that can return profit in a shorter period, so you need to determine what that profit value is today. If the future cash flow discount rate is 8 percent and the potential project return is $125,000, the present value can be calculated as $125,000 / 1.08 = $115,741.

NPV = Present Value (PV) – Initial Investment

NPV = $115,741 − $100,000 = $15,741

To calculate NPV for multiple time periods, you can simply extend the number of time periods to include the full duration of a project as shown:

$$NPV = \frac{F_1}{(1+r)^1} + \frac{F_2}{(1+r)^2} + \frac{F_3}{(1+r)^3}$$

You also can write this equation as

$$NPV = \sum_{t=1}^{n} \frac{F_1}{(1+r)^t}$$

Return on investment (ROI)—ROI is simply the expected profit from a project investment. As simple as this may seem, this is one of the biggest questions management and the stakeholders will have at the onset of a project. It also is a large factor in rough evaluation of project selection. The calculation for ROI is as follows:

$$ROI = \frac{\text{Total Project Investment}}{\text{Total Project Cost}}$$

Project Charter

During the initiating phase of a project, before the project has officially begun, work is done to understand the project objective, output deliverables, impact or benefit to the organization, stakeholders who will be involved, as well as associated cost and schedule requirements. After this information is gathered, organized, and analyzed, decisions need to be made regarding the approval of a project. For every project that has been selected for consideration, there needs to be an approval process that formally creates the project and allows work activities to commence; this is called the *project charter*.

Purpose of the Charter

The Project Management Institute defines a project charter as "a document issued by the project initiator or sponsor that formally authorizes the existence of a project and provides the project manager with the authority to apply organizational resources to project activities." Although information has been gathered to define project objectives and deliverables as well as other areas of related activities to the project, most of this work is performed behind the scenes, and most of the organization may not even know about the official project at this point.

The project charter allows stakeholders, who in most cases include customers, to agree on the project objective, deliverables, scope, costs, and time frame so the stakeholders can reach a consensus and decide that the project will commence. If everyone is in agreement that the project details outlined in the proposal are fair and equitable to all parties, an agreement is then made to officially start the project. It is important the initial stakeholders and/or management representatives make a clear statement to those in the organization who will be associated with this project that it has officially begun and a project manager has been selected to commence organizing project activities.

Structure of the Charter

The project charter is a document in which initial high-level data and information concerning a proposal of activities that will result in an output deliverable are compiled for analysis and approval. Initial information is typically gathered from a business need or a proposal from an external customer requirement. This information, in most cases, is at a high level and lacking specific detail, but is complete enough to provide a general scope of the proposed objective. In some cases, an external customer proposal may include specifications that may be very detailed, and the proposed deliverable can be well defined from this document. In either case, the project charter usually contains details defining only the deliverable at this initial stage and nothing that defines detailed cost structures or language outlining contractual agreements of any kind.

It should be noted that the creation of a project charter is not the creation of a legal contract, and the approval and authorization of the charter do not constitute a legal or binding agreement. The charter serves only to outline project details concerning the proposed deliverable so that both parties can be in agreement as to the objective of the project. After the charter is approved and a project manager is assigned, more detailed information of work activities, costs, and schedule requirements can be defined so that actual legal contracts can be drawn based on detailed information.

Because the project charter defines the official beginning of a project, some key elements are required to effectively facilitate the charter process. The following key elements should be included in a project charter:

- Project identification
- Sponsor and stakeholders
- Proposed start and completion dates
- General description, objectives, and requirements
- Milestones
- Assumptions and constraints
- Summary budget
- Summary of primary risks
- Project approval requirements
- Project manager and authority level assigned
- Individuals identified for charter approval

A project charter for a new product development is illustrated in Figure 2.9. This example shows the basic outline of suggested elements to be included in a project charter.

Project Charter			
Project Title:			Date:
Revision:	Project ID:	Prepared by:	
Sponsor Organization:		Sponsor Representative:	
Project Start Date:		Project Completion Date:	
Project Purpose			
Description:			
Objectives:			
Requirements:			
Project Schedule			
Milestone Schedule:			
Assumptions & Constraints			
Regulatory or Permits Requirements:			
Project Cost			
Estimated Project Budget:			
Payment Schedule:			
High-Level Risks:			
Project General Information			
Project Success Criteria:			
Project Manager Responsibilities:			
Stakeholders:			
Authorization			
Project Approval:		Project Approval Date:	

Figure 2.9 Project charter

Process or Artifact

As you have seen, the project charter plays an important role in the approval process of projects within an organization. All project proposals that are presented for consideration go through the project data-gathering, selection, and approval process. How this process plays out might be different depending on the organization. Some organizations with a more structured project management process and PMO may have a formal project charter process that is performed and documented in an actual project artifact called the project charter. The charter is a physical document, as described in this chapter, and is routed to the appropriate stakeholders for approval and signatures. This is the case for internal improvement projects and formal projects in conjunction with external customers.

Other organizations with a less formal project management structure go through the project *charter process* but might not use an actual written document for data-gathering and approval purposes. Because the project charter process outlines the activities required to properly document high-level information of a proposed project and serves to facilitate the approval process, some organizations carry out these functions using other documents and processes. It is important to note that no matter how the organization chooses to document the items performed in the project charter process, the charter function, as it is designed, still plays out as a requirement to officially approve and start a project. It is also important to note that whether the organization uses a formal physical document called the project charter or carries out charter functions through other means, the success of the project is directly correlated to the attention that is given to accurately completing all the elements in the charter process.

Review Questions

1. Explain how projects originate and why projects are used in an organization.
2. Discuss why managing stakeholder expectations is important.
3. What is the difference between product and project scope?
4. Explain why organizational constraints can be a concern to the project manager.
5. What is the drawback in using qualitative project selection techniques?
6. Discuss the purpose of a project charter and when is it used.

3

Planning Process

Introduction

Stakeholders and management tasked with information gathering, evaluation, selection, and approval of potential project opportunities must be aware of critical steps that must be completed before project planning can actually begin. In the initiating process, stakeholders generate two primary documents: the project charter and stakeholder registry. If a project manager was not a part of this initial stakeholder group, he is identified before project planning can begin. If the initial stakeholders have completed all elements of the initiating process, all information that was gathered and evaluated is given to the project manager to be used in developing a more detailed outline of project activities, costs, schedules, and deliverables. As previously mentioned, it is also vitally important that stakeholders have communicated to appropriate staff within the organization that a project has been approved and planning, scheduling, and procurements can begin for work activities.

The responsibility of developing the project plan typically is assigned to the project manager. Information that has been gathered for purposes of selection and approval is used in the initial creation of the project plan. Depending on the size of the organization, medium- to smaller-size projects may require only the project manager to develop the project plan, whereas larger size projects may require additional staff to assist the project manager in developing the project plan. It is advisable that no matter how large or small a project is, there is always one project manager responsible for the entire project. This helps to ensure that

- One person understands the scope and vision of the project objective.
- One point of contact is available for the customer as well as other staff within the organization to ensure optimum communication.
- One person reports on project status.

- One person is responsible for managing risk events.
- One person is responsible for delegating and scheduling project activities and procurements.

To help projects run smoothly, mitigate or eliminate risks, and ensure successful completion of the project objective, management staff must realize the importance of the project manager role. The manager assigned to oversee a project not only should be skilled in developing the project plan and managing resources and project activities, but also have project management knowledge of cost and schedule control, risk management, and communication and stakeholder management. After the project manager is selected and all information gathered during the initial process has been passed to the project manager, the management staff need to create a master plan of the project; this is called the *project management plan*.

Develop Project Management Plan

Information gathered in the initial phase of the project focused primarily on project deliverable details, general technology required, and capabilities of the organization to accomplish the project objective. Because these general areas of information were negotiated and agreed upon by the customer in the supplying organization, this step represents only the beginning stage in defining all the aspects of project management required to produce project deliverables. The first document that the project manager creates will address all the areas of project management that will be encountered throughout the project life cycle. This document is called the project management plan.

Project Management Plan Versus Project Plan

We must first address a critical area of terminology to clarify two terms that are commonly used out of context or used incorrectly to label a process of project management. It is important that project managers and other staff within the organization use project management terminology correctly so as not to misinterpret what project managers are intending to communicate. The following definitions help clarify two commonly misused terms in project management:

- **Project management plan**—This document explains how project processes are defined and implemented. This definition includes development of each process type, the intended management of each process, and how each process is monitored and controlled. Processes can include defining scope, schedule, cost, quality, human resources, communications, risk, procurements, and stakeholder management.

- **Project plan**—This definition identifies all work activities required to accomplish a project objective. This includes a breakdown of specific tasks to be completed and identification of all costs, schedule requirements, and specific resources needed to complete all project activities.

Project Management Plan Structure

As project management has developed and matured, various organizations have developed their own approach to what is required in structuring the master project management plan and, in some cases, an optimized plan based on a specific structure of the organization. In organizations where project management is not very well developed, project managers have taken upon themselves the need to craft their own project management plan; this may be good or bad, depending on the experience of the individual project manager. Figure 3.1 illustrates processes that can be included in a project management plan.

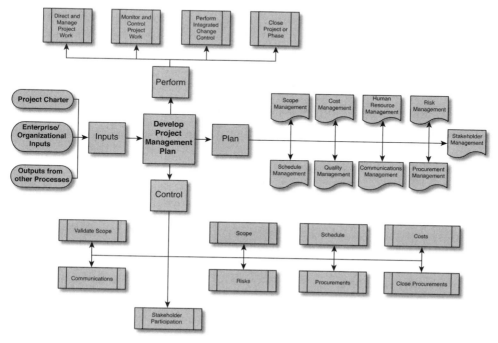

Figure 3.1 Project management plan

The project management plan has several areas detailing both inputs to the plan and outputs that will be created from processes within the plan. The plan is designed to be a global document that draws on inputs and information from the project charter and other areas in the organization with information pertinent to this project. It includes the process that defines all the work that will be performed in accomplishing the project objective as well as process development such as management and control of several areas within the project. Because this chapter is focused on planning a new project, we first address how each of the nine process areas is managed. It is important to note that these areas define how each process is intended to be *managed,* not necessarily outlining the details within the process itself. A brief explanation of how to develop a plan for managing each process is provided later.

- **Scope management**—Within the scope management plan, processes need to be established to manage how to define, validate, and control scope. It is important to manage what information is available to influence the scope, guidelines established to control the development of scope, and the information that the scope definition will provide to other processes in the project management plan. Managing the flow of information is the key component to developing scope management.
- **Schedule management**—Developing a plan to manage the schedule also includes developing processes that control and validate information used in developing the project schedule. A plan is needed to manage execution and control of the schedule during the course of the project. Especially on larger projects, it is important that the project manager have clear guidelines outlined in the schedule management plan to control how scheduling is implemented and how any change requirements need to be addressed.
- **Cost management**—Developing the plan to manage cost also includes processes to control the accuracy and validation of information gathering, manage influences within the organization in developing the budget, and control plans for how to manage cost throughout the project life cycle.
- **Quality management**—In some cases, developing plans to manage quality on a project can be as easy as analyzing measurable criteria, but in other projects, defining quality measurements and how quality plays out on the project can be more difficult due to the subjective nature of the project objective and deliverable. Planning quality, from a higher viewpoint, should always be about trying to measure completed work against a standard based on customer expectation. So planning quality should start with a clear understanding of the customer expectation and breaking down the deliverable into measurable components.

Project managers sometimes develop quality measurements of other areas on the project, such as human resources, procurements, engineering, manufacturing, and management. In the quality management plan, project managers generally outline what will be graded, and in the control quality process, they outline how they will be graded.

- **Human resource management**—In some cases, developing a plan to manage human resources within a project can be very simple, but in most cases, it ends up being a complex task requiring consideration of several areas of management. Planning human resource management should include, at a minimum, a project hierarchy, a responsibility assignment matrix, and a definition of authority matrix. Other areas that might need definition include how to manage external contracted staff, project staff located in other facilities or countries, and guidelines for disciplinary action if applicable.

- **Communications management**—Developing a plan to manage communications is one of the most important investments of the project manager's time. A common theme on most projects, which project managers reluctantly admit, is the number of problems that could have been avoided if there had been better communication. The project manager must understand the stakeholders', project team's, and upper management's information needs and requirements to design effective communication management within the project and the organization. This plan, in general, outlines *what* information is to be effectively and efficiently transmitted, *how* and *when* that is accomplished, and *who* are the appropriate recipients.

- **Risk management**—Planning for risks, ironically, is one of the most underdeveloped areas of a project from a planning standpoint. In most cases, project managers manage risk from a reactive standpoint rather than a proactive standpoint. Proactively identifying high-probability and high-impact risk events simply means identifying them and having a plan of action ready in case they occur. The project manager defines what tools can be put into place to conduct risk identification, analysis, prioritization, and contingency development and implementation. Planning risk management is a key element because it can both positively and negatively influence other processes such as cost, schedule, procurement, human resources, and other areas on the project.

- **Procurement management**—The procurement management plan primarily outlines processes for the make or buy and use of internally or externally contracted resource type decisions, supplier selection, and in some cases, first article validation processes; it also defines purchase and approval authority.

- **Stakeholder management**—The project manager must be aware that stakeholders are a vital component to the success of a project, so care must be taken in developing a plan to manage stakeholder expectations, information needs, and requirements, as well as their overall influence to the project. Stakeholders may be upper management, project team staff, and customers, each having different connections to the project and requiring individual management plans.

How to Use a Project Management Plan

The project management plan ends up being a compilation of several processes that the project manager uses to effectively and efficiently plan, organize, implement, and control all project activity. This plan is a tool developed by the project manager to document how project processes are to be conducted. It will serve as a project artifact, throughout the project life cycle, as a reminder to not only the project manager but also other project staff how processes were intended to be carried out. As the project progresses, modifications can be made to the project management plan to further clarify and define process detail. These changes should be conducted through the change management process.

After an outline of all components that will be identified within the project management plan is developed, it should be archived as a template to use on future projects. This is typically performed in the project management office (PMO), but if an organization does not have a PMO, it can simply be saved by a project manager to be distributed at the beginning of each project within the organization. This is a valuable tool in standardizing how projects are documented, planned, and controlled and can serve to mitigate or eliminate mistakes during the planning phase of the project.

In some cases, the project manager uses the project management plan to communicate to other project staff details of how processes were intended to be managed and controlled. The project manager may also elect to use the project management plan as a working tool among the project staff to maintain consistency while conducting processes throughout the project life cycle. In some organizations, senior management may require the project manager to submit a project management plan before activities can actually begin. This is to ensure that the project manager has planned management and identified and documented control of critical areas, and if appropriate, communicated to project staff to promote a successful project.

Collect Requirements

Now that a project plan has been developed and is in place to define all the processes that will be carried out during the project life cycle, it is time for the project manager to begin work on defining and planning work activities that will accomplish the project objective. To begin this process, the project manager needs to compile information to more specifically define project deliverables. This process is called *collecting requirements*. Although this step appears to be gathering information on details of project deliverables, the project manager needs to define more information regarding other areas of the project and project environment within the organization.

Definition of Requirements

In many cases, the area of collecting requirements is misunderstood as collecting all the requirements to specific project deliverables. Although specific details gathered will be in addition to what is stated in the project charter and statement of work documents, project requirements necessitate a broader view of the project for an organization. These higher-level requirements include

- **Requirements internal and external to the organization**—Internal needs or requirements can include alignment to the strategic objective of the organization; senior management's endorsement; functional management's cooperation; and human resources, equipment, and facilities. External requirements can include special conditions from the customer, availability of suppliers for critical items required, special contracted services, and governmental regulatory conditions that have to be met.
- **Stakeholder requirements**—These can also be internal or external to the organization but are more specific to individuals who have influence or direct responsibilities on the project. These requirements can be in the form of specific documents, analyses results, financial information, critical market timing data if applicable, and effective communication requirements of critical information.
- **Deliverable requirements**—These requirements include all detailed information defining specific requirements to project deliverables. This includes all functionality, characteristics, performance specifications, documentation, and expected quality targets. This can also include milestones for periodic inspections or customer approval.
- **Project requirements**—This information includes any special requirements that the project manager needs, such as project management software, special

project management staff to assist the project manager, any specialized professional analysis critical and required for project processes, or specific elements of the product. Depending on the complexity of the project, special financial analysis may be required for cash flow, multiyear net present value calculations, budget at completion analysis, specialized risk analysis, and payback or return on investment calculations. Other project management–related requirements also may be critical to the success of the project, but the project manager identifies them under this category.

Resources for Requirements

As you have seen, depending on the size and complexity of a project, several requirements may need to be identified during the requirements collection process. The scope of the requirements collection process is broad and can be either high level or very low level and specific. But where does this information actually come from? At this point in the project, the project manager must review what documentation has been created and retain sources for requirements information. The project manager can also seek out other resources internal and external to the organization to obtain requirements information. At this stage in the project development process, the following documents and areas in the organization should provide requirements information:

- **Project charter**—The project charter is used to outline a high-level description of the project objective and deliverable. Other details of the organization's capabilities, customer requirements, and expectations, as well as general high-level descriptions and definitions, can provide information for requirements collection.
- **Customer specifications**—If customers submit documents with specific definitions of product or service deliverables, they can be a great resource of very detailed information that can be used in requirements collection. Customer specifications can also be used to raise other questions of requirements such as special equipment, facilities, and human resources that also may be required.
- **Statement of work (SOW)**—This document might have been created during the charter process, if a written customer specification was not available, as a narrative description of the product or services that will be produced by the project. This statement might not be as detailed as a customer specification but will have enough detail to be used as a resource for requirements collection.

- **Stakeholder register**—The stakeholder register is created during the charter process. It outlines not only key stakeholders on the project, but also their needs and requirements as well as their relationship to the project and authority level on the project. This list of stakeholders can be a great source of information concerning several areas of the project; this can include project-specific knowledge, project or product knowledge that might be internal or external to the organization and market, or industry knowledge and experience.

- **Stakeholder management plan**—The stakeholder management plan is created as an output of the charter process; it contains other stakeholder information such as needs and requirements of information and communication, specific involvement with the project and management strategy of involvement, and details and management of stakeholder expectations. Although this document is designed more as an outline of how to manage stakeholders on a project, it can provide the project manager with information not found on the stakeholder register that can be used in requirements collection.

- **Subject matter expert (SME)**—If specific pieces of information cannot be gathered from either documentation or stakeholders on the project, the project manager can solicit the knowledge and experience of other subject matter experts within the organization. Subject matter experts are individuals who have either knowledge or experience of specific details pertaining to the project and/or product or service deliverables. It is usually advisable to seek out subject matter experts, if creditable and reliable, who can be either internal or external to the organization, for specific information because this will generally yield accurate and detailed information for requirements collection.

- **Historical data**—If the organization has a history of projects that have produced similar deliverables under like conditions, this is another area of valuable information for the project manager. Historical data can reveal details of project plans; work activities; cost and schedule outcomes; procurements activities; and use of human resources, equipment, and facilities. Historical data can reveal performance of not only employees but also external subcontractors who were used; this information can be valuable in defining the requirements of subcontractors for a new project. Historical data can also reveal lessons learned, which can be important in defining requirements and avoiding errors made in the past. Project managers can typically find lots of valuable information on past projects that will assist in requirements collection for new projects.

Requirements Management Plan

After the project manager successfully collects the intended requirements, it is advisable to document conversations and other information collected and then organize and archive all requirements documentation. The project manager then outlines a requirements management plan that defines how requirements information and activities will be planned, implemented, and monitored. The plan also includes human resources assigned to requirements activities, expected outcomes, cost, schedule, and prioritization of activities. This plan provides the project manager a roadmap that can be communicated to ensure critical requirements are completed accurately and on schedule. This plan can also include definitions of how requirements can be changed through a change management process and therefore documented, controlled, and communicated to the appropriate project staff and stakeholders.

Define Scope

After a project has officially begun and stakeholders and project management staff have gathered additional information and requirements, it is now time for the project manager to begin organizing information into an actual working plan detailing how project activities will be conducted and controlled. The project manager should have a great deal of information, having completed the project charter, which includes high-level information regarding project objectives as well as specific information from customer specifications. The project manager also should have gathered and organized the statement of work (SOW) and additional information so that he, project staff, and stakeholders have a clear understanding as to the scope of the project objective. To accomplish this, the project manager must be clear as to what it means to define scope.

Scope defines the boundaries containing all actions or functions that have been defined by either customer or organizational expectations of the project objective. If a project has been created out of the need for a product or service, two things are actually created: (1) the *product* or *service* itself and (2) the *project* that will contain all the activities to produce the product or service. So the project manager actually has two areas of scope to define: *product scope* and *project scope*.

Product or Project Scope

It is advisable for project managers to make a clear distinction between what they are trying to put boundaries around so that project staff and stakeholders can understand what is being defined and controlled. The project manager is responsible for controlling the project deliverable (specific product or service) in addition to all activities and processes associated with the project. The important aspect with defining scope is the perspective of stakeholder expectations and understanding the extent or range of work required to fulfill those expectations. The project manager must then define the boundaries and limitations to the product or service being developed and the amount of work performed within the processes of the project life cycle.

- **Product scope**—This is all required specifications defining a deliverable. Scope defining a product is specific in nature and usually results from information derived from customer specifications, statement of work (SOW), industry standards, or government regulatory requirements. The perspective of product scope is to put in place boundaries that ensure all the customer expectations of the deliverable have been met as well as control the extent of work performed on the deliverable to protect the supplying organization from overdeveloping a product or service that was not intended.

- **Project scope**—This is all processes required to define, implement, monitor, control, and close project activities. Scope defining a project can be both specific and generalized in nature. The project will have specific work activities and processes as well as a general nature of project management with regards to stakeholders and environmental factors. The project manager must set boundaries to clarify not only what is being produced by the project, but also who will be involved in the project. He also must maintain the size and complexity of the project and manage project procurements, use of facilities and equipment, and human resource and stakeholder management. Depending on the structure of the organization and the authority level of the project manager, project scope management is one of the most important functions of the project manager so that he can effectively manage cost, schedule, and quality of the deliverable or service required by the project.

Who Defines Scope?

Now that you know two types of scope must be defined, this can raise the question as to *who* is responsible for defining product scope as well as project scope. *Product scope* is usually outlined in initial documents such as the project charter and statement of work, but is usually further defined, before the project begins, by customer specifications, contracts, and other documents with specific information relating to the product or service. The majority of this information is provided by the customer and/or negotiated between the supplier organization and customer. *Project scope* is usually a much larger-scale compilation of project documentation, environmental factors, organizational processes, and all other documentation required to establish boundaries around the processes performed during the project life cycle. Depending on the size and complexity of the project in the structure of the organization, it is generally understood that the project manager holds the responsibility for developing the project scope. On larger, more complex projects, the project manager might solicit the help of other project staff in defining scope for various processes due to the enormous amount of information that has to be analyzed and processed to correctly define scope. The project manager ultimately holds the responsibility of managing both the product scope and project scope.

Project Scope Statement

Because a great deal of information has been gathered on details outlining the scope for both the product and the project, the majority of this information is organized into a document called the *project scope statement.* The project scope statement should contain, at a minimum, the following project information:

- Project objectives
- Project deliverables
- Product scope
- Work breakdown structure (WBS)
- Acceptance criteria
- Scope validation and control processes
- Special exclusions
- Regulatory requirements
- Internal or external restrictions or limitations
- General description of high-level risk
- Assumptions

The project scope statement can also be used to form the general baseline used to create the project plan of activities and guidelines to establish boundaries for work activities. This document may have a combination of descriptive text as well as tables, graphs, and charts that may assist in communicating details of project scope. This document also needs to be developed in a format that can be easily understood by all stakeholders of the project. An example of a project scope statement is shown in Figure 3.2.

Project Scope Statement				
Project Title:				Date:
Revision:	Control Number:		Project Manager:	
Project Overview				
	1.	Objectives:		
	2.	Deliverables:		
	3.	Define Project Scope:		
Work Activities				
	1.	Create WBS		
	2.	Establish Project Baseline		
Clarifications				
	1.	Regulatory Requirements:		
	2.	Assumptions:		
	3.	Restrictions/ Limitations:		
Acceptance Criteria				
	1.	Required Inspections:		
	2.	Testing Standards:		
	3.	Define Pass/ Fail Criteria:		
	4.	Scope Validation Process:		
	5.	Scope Control Process:		
Major Risks				
	6.	Identify Major Risks:		
	7.	Scope of Contingencies:		
	8.	Define Triggers/ Activators:		

Figure 3.2 Project scope statement

Scope Management

The project scope statement is an important document that the project manager uses to control scope of a project. As the project begins and flows through the project life cycle, the project manager needs to monitor all work activities and processes to ensure work is being performed within the boundaries outlined within the scope statement. Project managers also use the scope statement as a baseline while processing change requests or in the evaluation of risk elimination, mitigation, contingency, or corrective action activities. Depending on the information and communication structure of a project, some change requests may be informally introduced; because they are not controlled, they result in extra work activity that may or may not be approved. This situation is called *scope creep*.

An example of scope creep might be a representative of the customer contacting someone on the project staff to change a feature on the deliverable. If this change is not managed through a change control process where the project manager is notified, appropriate stakeholders or project staff can evaluate the change and provide recommendations. Then these changes can result in increased cost, schedule delays, and unapproved alterations to the project deliverable. After a project has begun, changes are inevitable, and the project staff should welcome evaluation of changes but control them to protect both the customer and the supplier organization. The project manager uses the scope statement as well as the change control process to properly manage changes on the project. Change requirements may be in the form of product or service modifications, but may also come from other areas internal or external to the organization; they include

- Documentation updates
- Project management plan updates
- Work performance information
- Change requests through the change control process
- Organizational process updates
- Regulatory requirements

Projects have several areas of information input that can influence not only the product or service being developed, but also the processes carried out to produce the project deliverable. It is incumbent on the project manager to ensure that an effective and efficient change control process is in place and all stakeholders, project staff, suppliers, and customers are aware of the importance of managing changes through the change control process. Project managers must understand the importance of developing the scope statement and an effective scope management plan as the first step in developing project controls in the early stage of project planning.

Work Breakdown Structure (WBS)

The project manager now has the general idea of the project outlined by the project charter and statement of work, detailed information outlining project deliverables and project processes, and information defining the scope of the project as well as the product or service being produced. We now turn our focus to the work activities, resources, cost, and schedules required to produce the product or service required to complete the project objective. Depending on the size and complexity of the project and product or service that will be developed, this can be a daunting task to understand and organize all the work required to produce the product deliverables.

The process of creating the work breakdown structure starts with understanding how to break up a top-level project idea (product or service) into smaller components correctly and how far these components can be broken down to understand the actual work activity at the work package level. It is advisable that the project manager always seek the knowledge and experience of subject matter experts to understand how a product or service can be broken down correctly into smaller components. In some cases, project managers having knowledge and experience within a specific industry may understand how to initially divide up work packages. In other cases where project managers may not have specific knowledge of the product, the advice of subject matter experts can reveal the correct decomposition of a product or service into smaller subcomponents and eventually to the work package level.

Subdividing these components is critically important because this defines the work required and also forms the initial foundation for the project plan of activities, which provides detail in cost and schedule estimating as well as human resource, facility, and equipment requirements. If the subdivision of these components is not done correctly, project cost and schedule estimating will be incorrect, and will incorrectly define the project deliverable, resulting in failure to effectively complete the project objective and creating a negative impact to the customer and supplier organization. To assist in correctly and completely breaking down a project deliverable, the project manager can use a powerful tool called the *activity decomposition decision tree*.

Activity Decomposition Decision Tree

Breaking down a large item into smaller pieces is like reverse engineering or understanding the relationships regarding which pieces fit together. This task can be very difficult because one would have to visualize a completed item and then take it apart in his mind. Because most projects start with a defined objective or deliverable, understanding how it is built can be next to impossible. The project manager can use the activity decomposition decision tree to analyze how a finished deliverable is taken apart and broken down into its smallest components.

The decision tree analysis starts with a component, such as the final deliverable, and asks whether or not it can be broken up. If yes, then how many pieces and what are they labeled? If no, then a piece is considered the smallest component (work package activity) and the chain stops, as shown in Figure 3.3.

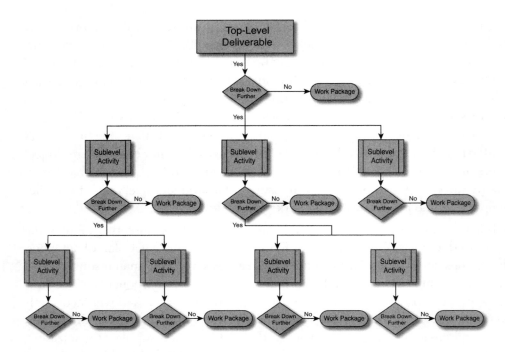

Figure 3.3 Activity decomposition decision tree process

As the components are broken down further, branches of work appear, and they are used to define various sublevels of work as described in the WBS. The process continues until all branches of decomposition have gone to the lowest-level work package, as shown in Figure 3.4.

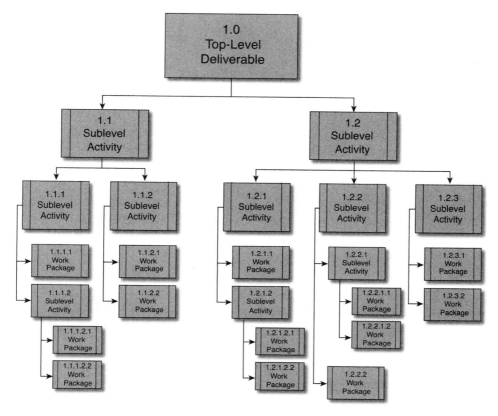

Figure 3.4 Activity decomposition decision tree to lowest level

Create the Work Breakdown Structure

To create a work breakdown structure, the project manager must ensure certain processes have been completed and documentation has been created as a prerequisite to organizing work activities in the WBS. The project manager uses the following documents as input information:

- Project charter
- Customer specification
- Statement of work
- Requirements documentation
- Project scope statement

The next step is the process of creating the master outline. For simple projects requiring only a few process steps, this can be done by hand on a piece of paper. For more complex projects, however, the use of well-known software such as Microsoft Excel or Microsoft Project is preferred to efficiently manage the detail of all information that will be organized. As work is being broken down into smaller components, it's important to try to understand the lowest level of work possible to get the highest detail for cost, schedule, and resource planning. It is also important to note whether one work activity cannot be started until a previous activity has been completed in this type of detail; if so, that will create *predecessor* or *successor* requirements. It is important in the organization of work activity to understand predecessor or successor requirements because this can impose limiting factors in the completion of higher-level components of work; this issue is covered in more detail in Chapter 5, "Activity Sequencing." Figure 3.5 uses MS Excel to illustrate the basic decomposition structure of a WBS and includes details of required levels as well as correct labeling of levels (based on information in Figure 3.4).

Task	WBS Code	Project Tasks	Durations	Predecessor	Resources
1	1	PROJECT NAME	74 Days Total		
2	1.1	First Sublevel Activity	29 Days SubTotal		
3	1.1.1	Lower divided Sublevel Activity	13 Days		
4	1.1.1.1	Lowest Level Work Package	8 Days		Name
5	1.1.1.2	Lower divided Sublevel Activity	5 Days		
6	1.1.1.2.1	Lowest Level Work Package	3 Days	4	Name
7	1.1.1.2.2	Lowest Level Work Package	2 Days	6	Name
8	1.1.2	Lower divided Sublevel Activity	16 Days		
9	1.1.2.1	Lowest Level Work Package	9 Days	7	Name
10	1.1.2.2	Lowest Level Work Package	7 Days	9	Name
11	1.2	Second Subtask	45 Days SubTotal		
12	1.2.1	Lower divided Sublevel Activity	5 Days		
13	1.2.1.1	Lowest Level Work Package	2 Days	10	Name
14	1.2.1.2	Lower divided Sublevel Activity	3 Days		
15	1.2.1.2.1	Lowest Level Work Package	1 Day	13	Name
16	1.2.1.2.2	Lowest Level Work Package	2 Days	15	Name
17	1.2.2	Lower divided Sublevel Activity	30 Days		
18	1.2.2.1	Lower divided Sublevel Activity	22 Days		
19	1.2.2.1.1	Lowest Level Work Package	10 Days	16	Name
20	1.2.2.1.2	Lowest Level Work Package	12 Days	19	Name
21	1.2.2.2	Lowest Level Work Package	8 Days	20	Name
22	1.2.3	Lower divided Sublevel Activity	10 Days		
23	1.2.3.1	Lowest Level Work Package	7 Days	21	Name
24	1.2.3.2	Lowest Level Work Package	3 Days	23	Name

Figure 3.5 Basic work breakdown structure in MS Excel

As shown in Figure 3.5, a hierarchy numbering system defines each level of work activity. This numbering system starts at the highest level (level 1.0), typically identifying the project title or top-level component, and continues down in this example using the decimal equivalent of lower-level components such as level-two work activities (1.1, 1.2, 1.3, and so on). It continues down to level-three work activities (1.1.1, 1.2.1, 1.3.1, and so on) and then down to level-four work activities (1.1.1.1, 1.2.1.1, 1.3.1.1, and so on) until reaching the lowest work package activity. This decimal breakdown methodology is normally used to identify level, as shown in MS Project in Figure 3.6.

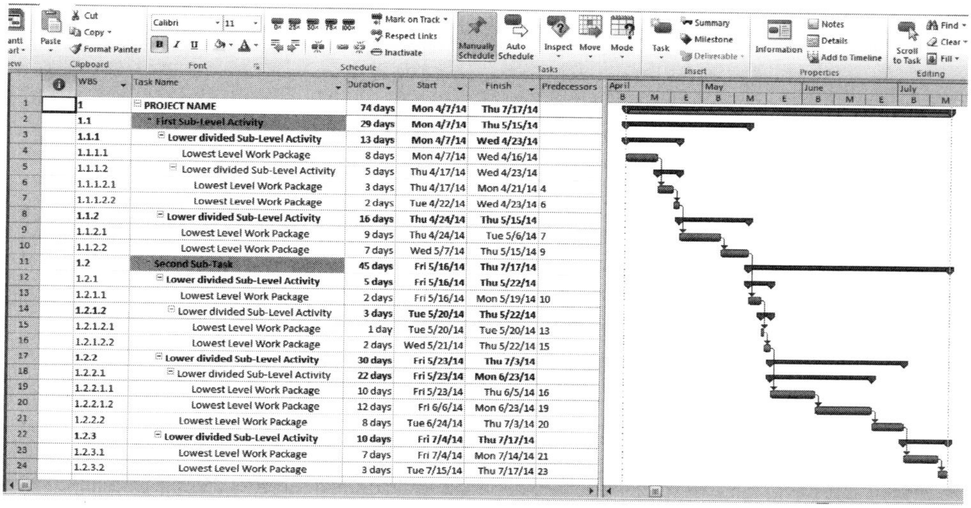

Figure 3.6 Work breakdown structure in MS Project

The project manager takes several steps in defining cost, schedule requirements, and resource allocation for work activities; they are covered in later chapters. After developing the basic outline, the project manager must review the breakdown of work activities to ensure they represent the lowest level of work required, that this decomposition is accurate and correct, and that any sequential order has been verified along with predecessor or successor requirements. This is a critical step in project development. Investing quality time to review the details in the WBS is advised because the rest of the project will be based on this structure.

Review Questions

1. Discuss how the project manager would use a project management plan.
2. Explain the difference between product scope and project scope.
3. Discuss how the project manager would manage scope on a project.
4. What is the primary function of the activity decomposition decision tree?
5. Why is it important to reduce deliverables to a smallest component?

Part 2

Project Schedule Analysis

Chapter 4 Activity Definition 79

Chapter 5 Activity Sequencing 97

Chapter 6 Resource Estimating 117

Chapter 7 Activity Duration Estimating. . . . 139

Chapter 8 Schedule Development 157

Development of the project schedule falls within the planning phase of the project life cycle. At this point in the development of a new project, stakeholders and the project manager have enough information to understand the general idea of the project objective and any required project deliverables. The project manager and any accompanying project staff now begin the process of acquiring details that will allow the project manager to outline specific work required; analyze scheduling for human resources, facilities, and equipment; purchase materials; and schedule milestones, inspections, and project deliverable dates. The chapters in Part 2 outline processes and tools specific to activity definition and sequencing, resource and duration estimating, and development of the master schedule.

Plan Schedule Management

One primary output of a project's planning phase is the master schedule of all work required to complete the project objective. With most projects, either large or small, creating this schedule requires a great amount of work and organization of information. Project managers must always keep in mind that organization is the top priority in their approach to developing a project and effectively managing it. Organization only comes through planning and documenting information. Defining activities results in an enormous amount of information and details that need to be captured and organized to effectively create a project schedule; this task is accomplished through a *schedule management plan*.

Planning schedule management is not the creation of the schedule itself, but the project managers' initial thoughts regarding what tools will be used to gather and document information, define work activities, and develop a project schedule. Depending on the size of the organization and the size and complexity of the project, project managers may perform these actions independently using their own methodologies. In some cases, the project management office (PMO) may have a standardized process that all project managers in the organization use in developing an initial schedule. Some organizations organize a small team of managers, including the project manager, to define initial project structuring and reporting protocols. In any case, planning schedule management should include, at a minimum, the following:

- Project management tools used to break down and organize work activities
- Key people to interview for information
- Ways to accurately and effectively document information gathered
- Units of measure used to document the project (that is, hours, weeks, months)
- Scheduling estimating techniques
- Scheduling methodologies
- Tools used to monitor and measure project performance
- Reporting formats and protocols

After making decisions on the preceding items, project managers can use this document as a standardized guideline to begin gathering and documenting information as well as structuring a project schedule. This document also can be used to define the schedule planning process for future projects within the organization. It is also important that individuals within the organization tasked with project management responsibilities understand the importance of standardizing these processes across the organization. With the development of these processes comes their review and constant improvement, which helps project managers efficiently develop accurate and effective project scheduling and schedule management techniques.

4

Activity Definition

Introduction

As projects are created and information is gathered to define the overall objective, this process usually results in the definition of a project deliverable but rarely defines the levels of detail required to produce that deliverable. The next step the project manager takes in developing the project plan is to subdivide one project deliverable into several subcomponents of actual work to be completed. This subdivision of work activities to the smallest components is required so that as much detail as possible can be gathered to define the characteristics of each work activity.

The project manager spends much of his time managing cost and resources and scheduling day-to-day work performed on the project, and this work can be accomplished effectively only by having enough detail regarding those work activities to understand what has to be accomplished. It is also important the project manager develop a detailed plan showing the sequence of work activities and resources assigned to each activity to effectively manage the allocation of resources and overall creation of a project deliverable. Work activities also include the assignment of human resources at various levels of responsibility. The project manager has to develop a plan to organize information regarding human resources, their connection to the project and work activities, and authority level. The project manager typically uses an organizational tool to outline the responsibilities of human resources and to communicate to stakeholders and project staff scheduled work activities.

Activity definition is the most critical process of the project life cycle because it not only defines the details of project deliverables, but also yields definitions in the structure of all work activities required to meet project objectives. It is important the project manager and project staff allocate sufficient time and resources to define details of work activities and properly characterize each work activity so project planning can be comprehensive, effective, and efficient. This chapter outlines several tools used to define work activities, document and track human resource responsibilities, and organize work activities into a sequential schedule that can be used to manage the completion of project deliverables.

Activity Analysis

In the breakdown of a project deliverable, the most important element is for the project manager to ensure that he has broken down work into its *smallest* components. These components, called *work packages,* are what the project manager uses to define details required for each component of work. It is now time to perform an analysis of individual work package activity to determine critical pieces of information the project manager needs to build the project schedule and budget.

It is critical that project managers develop tools to help ensure processes are performed correctly, completely, and consistently. Project managers should analyze work activities on every project because critical information comes from this analysis. One process tool covered in Chapter 3, "Planning Process," was the *activity decomposition decision tree.* This process is designed to provide a systematic breakdown of a project deliverable into its smallest components and should be the first step in activity definition: defining the smallest activity.

The second tool project managers can use is the *activity information checklist* (see Figure 4.1). Project managers have a great deal of responsibility to the information they are managing on a daily basis, and having tools that simply remind them what they're supposed to do is valuable in a hectic environment such as project management. The activity information checklist is simply a reminder of what information will be gathered for each work activity so the project managers have all the information needed to properly define, sequence, schedule, and budget work activities. Although project managers are encouraged to develop tools like this on their own, the activity information checklist should include, at a minimum, the following information:

- **Identifier**—A label or name for the work package
- **Description**—A brief description of activity to be performed
- **Material required**—Any material that needs to be purchased for this activity
- **Work activity**—Actual work to be performed, such as "write software code," "pour concrete," "design printed circuit board," or "design the machine housing"
- **Human resources**—Human resources required to perform the activity

- **Equipment/facilities**—Any equipment required and any special use of facility space
- **Time duration**—Estimate of time required to complete the activity
- **Activity cost**—Estimate of total cost to complete activity
- **Predecessors**—Any project work activities that have to be completed before this activity can begin
- **Risks**—Any primary risks that could have a significant impact on the project budget or schedule

Activity Information Checklist	
Project Name:	
Name: (person collecting information)	Date:
Title of Work Activity:	
WBS Code:	Activity Revision:
Description of Deliverable:	
Required Work Activities:	
1	
2	
3	
4	
Required Materials:	
1	
2	
3	
4	
Required Human Resources:	
1	
2	
3	
4	
Required Equipment/ Facilities:	
1	
2	
3	
4	
Total Cost:	Total Time Duration:
Major Risks:	Predecessor Activities:
1	1
2	2
3	3

Figure 4.1 Activity information checklist

Using a simple checklist ensures the project manager does not forget critical pieces of information and helps build consistency in the information-gathering process for all work activities on the project and for other projects within the organization. Now that the project manager has a simple checklist tool to use for information gathering on work packages, the next process he must perform is gathering this information.

Activity Information Gathering

Using a simple tool such as the activity information checklist helps give the project manager a direction for what information needs to be gathered for each work package. Project managers and project staff identified to assist the project manager in gathering information must be aware that this process, at the beginning of a project, can be one of the most crucial and should be taken seriously. These individuals should be qualified to gather information and should know the importance of correct, complete, and accurate information and what impact it can have on the project. To perform this process correctly, the project manager and/or stakeholders need to consider three primary components for successful information gathering:

> **Who is gathering information?** The first component is selecting individuals who have knowledge, background, and experience in the work package activity for which they are gathering information. This step is critical in information gathering because these individuals need to be able to decipher what information is important and how much information to obtain, and to verify information accuracy. These individuals also should know the best sources for reliable and accurate first-hand information. The project manager must understand the critical nature of having competent staff in performing this process and the ultimate impact this process has on the overall project.
>
> **Where does the information come from?** The next critical component is knowing where information is gathered from because the accuracy and completeness of this information forms the foundation for developing the project cost, schedule, and specifics on defining project deliverables. All the information required to define work activities is available somewhere in some way, shape, or form, but it is the individuals selected to gather this information who must know where and how reliable and accurate information can be obtained. Individuals qualified in gathering information for work activities should know the value of first-hand information, even if it is more difficult to obtain, as this usually holds a higher reliability factor than second- or third-hand information. Other nonhuman sources of information, such as archived information within the organization and Internet sources, should also undergo the same scrutiny by

those qualified to gather information. The emphasis should first lie in qualifying the reliability of the source and then determining the accuracy and completeness of the data. Examples of more reliable information sources are as follows:

- Customer specification
- Project statement of work (SOW)
- Scope statement
- Subject matter experts (SME)
- Organization's database (archived projects and "lessons learned" documents)
- State and local regulatory documentation

How accurate and complete is the information? The third critical element of information gathering is the information itself and how relevant, accurate, and complete the information is for correctly defining the work package activity. The project manager should understand that the information gathered for a work package is the product of both the person gathering the information and the source of the information. It is ultimately the responsibility of the individual gathering information to determine when sufficient information has been gathered and that this person is comfortable with the reliability of the source in providing accurate and complete information.

Activity Organization

As the project manager begins to compile information on specific work packages, he needs to organize it in a form that he, stakeholders, and the project team can view and/or use for project development, project management, and communication of details and work assignments for project activities. As discussed in Chapter 3, after project deliverables are broken down into their smallest components and information is gathered to define specific work package activities, the project manager can use a tool such as the *work breakdown structure* (WBS) to organize specific work packages and document work activity information.

Organizing project activities involves not only defining and documenting where specific work activity belongs within a project structure, but also organizing details and periphery items relative to a specific work package activity. The project manager also needs a definition outlining the individual responsible for that work activity, and if the project manager is to oversee that work or there will be a responsible individual overseeing a team of individuals performing work for that activity. Projects may have work packages with individual contributors that are easily managed, whereas other projects might have work packages with several human resources and materials and equipment required to perform that activity. Activity organization then has to be designed at two levels: work package activity level and overall project level.

Activity level—In some cases, a specific work package activity might have only one human resource performing one work activity with minimal materials or equipment required. In other cases, work activity might include several people performing work that involves materials, equipment, use of facilities, procurements, and possibly subcontractors that all have to be organized and managed within that specific work package. The project manager must understand the scope of work required in each work package so as to organize all resources, materials, and equipment so that work can be completed correctly and efficiently. The project manager might have to coordinate actions from several departments within the organization and possibly external resource requirements that have to be organized so work activities move seamlessly with minimal or no scheduling conflicts.

Project level—The project manager using an organizational tool such as a work breakdown structure must organize work activities in sequential order so as to complete work packages in the correct order to complete higher-level components. As depicted in the decomposition decision tree, smaller work is completed first and joined together to build larger components of the project deliverable. The project manager has to identify the order of work packages and develop a seamless sequencing of work that will progressively produce larger and larger components of the project deliverable.

The project manager also has to identify and coordinate all periphery activities to work packages such as procurements, delivery schedules, scheduling of equipment and facilities, scheduling of human resources available within the organization, and the contracting of external resources if required. Because some of these resources require contract negotiations, these activities have to begin far in advance of their need on the project to ensure all details have been ironed out prior to their requirement. The area of procurements typically struggles with items such as stock availability, long lead times, and challenges with deliveries staying on schedule. The work breakdown structure assists the project manager in giving a visual representation of all work activities required on the project organized in a sequential timeline. Consequently, the project manager can keep track of critical tasks that are time sensitive to keep project activities on budget and on schedule.

Activity Definitions in the WBS

The project manager uses the work breakdown structure initially to organize all work activities broken down to their smallest components. Because this tool serves as a great visual representation of the sequence of work required to produce a project deliverable, this tool allows the project manager to see a bigger picture of all periphery items that are required throughout the project life cycle to support work activities being accomplished. As stated previously, many of these items need to be addressed out of sequence from when they are supposed to happen, so the project manager needs to know when critical periphery items have to be addressed so their required contribution will be ready when required on a work activity. Figure 4.2 shows a sample work breakdown structure with project activities and periphery items that the project manager can see far in advance of their need to be addressed.

1.0		Build House
1.1		Initial Work
	1.1.1	Develop Plans
	1.1.2	Get Permits
	1.1.3	Secure Funding
1.2		Foundation
	1.2.1	Level Ground
	1.2.2	Foundation Markers
	1.2.3	Dig Ditches
	1.2.4	Install Forms
	1.2.5	Install Sub-Plumbing
	1.2.6	Install Sub-Electrical
	1.2.7	Install Rebar
	1.2.8	Inspection
	1.2.9	Pour Footings
	1.2.10	Pour Slab Foundation
1.3		Framing
	1.3.1	Frame Walls
	1.3.2	Install Roof Trusses
1.4		Rough Electrical
	1.4.1	Install Main Panel
	1.4.2	Set Boxes
	1.4.3	Pull Wire
1.5		Rough Plumbing
	1.5.1	Install Sewer Drains & Vents
	1.5.2	Install Copper Lines

Figure 4.2 WBS with activities and periphery items

When project work activities are arranged in the work breakdown structure and the sequence of work activities is developed, information and details that have been gathered on each work activity can now be documented. Information of a work activity can be documented in several different ways. This discussion covers two simple tools that project managers can use: Microsoft Excel and Microsoft Project.

The first example uses Microsoft Excel in a simple work breakdown structure, giving two ways that information can be documented for each activity. This can be in the form of a note attached to the work activity if only a minimum amount of information is being documented. If more information such as customer specifications, drawings, and other more detailed information must be listed for this activity, a hyperlink can be used to another workbook page that will have more room for this type of information about that activity, as shown in Figure 4.3.

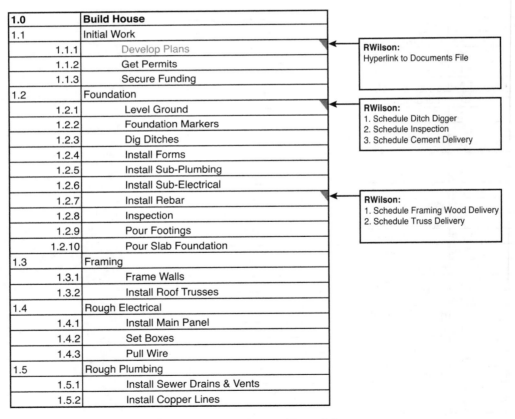

Figure 4.3 WBS with activities, comment notes, and hyperlink

Another tool project managers can use is Microsoft Project. In this example, a simple work breakdown structure is illustrated, with corresponding activity information locations within Microsoft Project where detailed information can be documented. Microsoft Project can be used to manage projects because it not only documents activities and the background information for these activities, but can easily facilitate project monitoring and control as well as display graphs, charts, and reporting features. If the project manager does not already know whether the organization uses this software, it is best to inquire whether the organization has already purchased software that includes project management features. This tool might be a way to standardize project management documentation within the organization. If the organization does not have existing software that can perform these functions, independent software packages such as Microsoft Project are inexpensive and easy to obtain.

In the example shown in Figure 4.4, Microsoft Project shows different areas of information that can document information for work activities on a project. Some of these areas, such as the Task Information window, can include very detailed information regarding human resources, external subcontractors, predecessors, and special notes, for example.

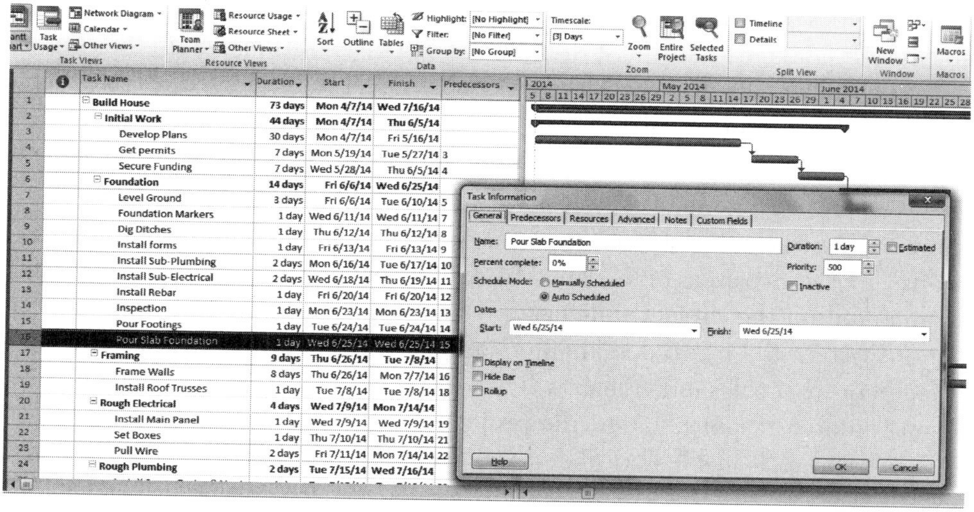

Figure 4.4 WBS with activities using MS Project and captions

The important point with gathering and documenting information is that the more information the project manager has for each work activity, the more confidence the project team will have that they are performing activities correctly and completely. Documenting project activity information in an organized and efficient form also gives the project manager a tool for monitoring and controlling work activities to ensure they stay on schedule and on budget. Having activity information documented in a form that can be archived within the organization is also helpful for new projects; this information can be accessed to confirm new information that is gathered for activities or can be analyzed from a "lessons learned" standpoint.

Responsibility Assignment

One important component of the information-gathering process for a project is defining and clarifying the responsibilities of all project staff and stakeholders who will be involved with the project. Because human resources are being identified for certain work activities, it is important that the responsibility assigned to each human resource be documented and effectively communicated. Project managers spend a great deal of time in their day-to-day activities managing human resources. Consequently, having a tool to consolidate what responsibilities have been assigned to certain individuals is valuable in managing these resources.

Different types of human resources are involved with a project, such as stakeholders, project staff, and customers, but these labels can be confusing within the project because some individuals may appear to fall under several different titles. Having a tool that categorizes human resources, their role on a project, their authority level, and the specific work activity assigned to them is a necessity for not only the project manager, but also all who are involved with the project.

Before we discuss tools that are used to organize human resource responsibilities, we must first look at two areas of information about resources that have been gathered and documented on the project thus far: the stakeholder registry and work activity information. Within these two documented areas, the project manager has information to decipher what roles individuals will have on the project. To assist in categorizing human resource responsibilities, the project manager first needs to define what human resources are available and what responsibilities they will have.

Responsibility Definitions

In gathering information about human resources required for a project, project managers will find it necessary to categorize those resources either by general involvement and responsibility or by specific work activity. The first-cut in categorizing is to define direct or indirect involvement. Examples of *directly involved human resources* are

- Project manager
- Project team (all individuals with specific work activity assignments)
- Stakeholders with direct involvement
- External human resources with work activity assignments

Some human resources will not actually have work activities but will be involved on a project. They include management staff, executives, and those involved with providing or requiring information from the project. The project will also have many supporting staff internal and external to the organization that might *not have direct and specific work activities* but may perform supporting functions. They include

- Quality assurance
- Manufacturing engineering
- Design engineering
- Administrative, accounting, and human resources departments
- Shipping and receiving
- Legal advisory and contract negotiating roles
- Functional managers, supervisors, or leads who can provide direction in mentoring and training and additional information for project activities

Because several types of resources are involved in most organizations, the project manager must manage these resources during the project life cycle and in doing so must have a way to organize what resources are required and what types of resources they are. The next step in organizing human resources with direct involvement responsibilities is to create a general category describing what functions human resources play on a project. Figure 4.5 shows a *resource assessment* for a Telecom project where the project manager can define what general types of resources will be used.

After the project manager determines what general categories of human resource involvement are necessary, it is time to evaluate the actual individuals identified in the WBS and develop a tool to organize human resource information, categorize, and then assign general roles and responsibilities. This task can be accomplished using a *responsibility assignment matrix.*

Resource Type	Quantity Required	Duration			Skill Level				Employment Type			
		Single Activity	Multiple Activities	Entire Project	Entry/Intern	Mid Level	Expert	Managerial	In-House	Contracted	Scheduled in Advance	On-Call/As Needed
Electrical Engineering Lead	1			X			X	X	X		X	
Electrical Design Engineer	4		X			X	X		X		X	
Mechanical Design Engineer	2		X			X	X		X		X	
Software Engineer	3		X			X			X		X	
Engineering Technician	2			X			X		X		X	
Engineering Assembler	2		X			X			X		X	X
Test Rack Engineer	1		X				X		X		X	
Test Technician	2			X		X			X		X	
Incoming Inspection Officer	1		X			X			X			X
Packaging Engineer	1	X				X				X		X
Quality Engineer	1		X				X		X			X
New Product Introduction Engineer	1		X			X			X			X
Marketing Manager	1		X			X		X	X			X
Sales Manager	1		X			X		X	X			X

Figure 4.5 Resource assessment

Responsibility Assignment Matrices

The best way to document and organize human resources and their responsibilities is to use a tool called a responsibility assignment matrix (RAM). Each project manager can develop this tool uniquely depending on the size of the project and the variety of human resources that will be involved. Some organizations that have a PMO may have a standardized assignment matrix template the project managers can use. This tool is used to clarify who has what type of responsibility, by using letters such as **P** = Primary responsibility and **S** = Supporting role.

Another similar tool is the *responsibility, accountability, consultative, and informative* (RACI) *matrix*. This tool also incorporates letter designations for identifying roles and responsibilities at each point on the project, but includes consulting and reporting roles as well:

R = Responsible; person assigned to a work activity (person doing the work)

A = Accountable; for verification of work activity completion, sign-off, validation

C = Consultative; subject matter expert or stakeholder who can be solicited for information or decision making

I = Informative; project staff, stakeholders, or management staff who need to be informed on work activity status

An example of this simple tool is shown in Figure 4.6.

RACI Matrix		Human Resources							
WBS	Work Activity	Owner/ Sponsor	Project Manager	Accounting Manager	Sales Manager	Facilities Manager	Purchasing Manager	Inventory Control Manager	Contractor
1.1	Business Plan								
1.1.1	Conceptual Design	R	A	C	C	I	I	I	
1.1.2	Business Structure	R	A	C	C	I	I	I	
1.1.3	Develop Plan	R	A	C	C	I	I	I	
1.2	Select Location								
1.2.1	Needs Assessment	R	A	I	C	C		C	
1.2.2	Evaluate Locations	C	A	C	C	R		I	
1.2.3	Final Selection	R	A	I	I	C		I	
1.3	Build Out Facility								
1.3.1	Interview Contractors	C	A	I	I	R	I	I	C
1.3.2	Get Building Permit	I	A	I	I	C	I	I	R
1.3.3	Construction Build-Out	C	A	I	I	C	I	C	R
1.4	Acquire Inventory								
1.4.1	Identify Inventory	C	A	I	R	I	I	C	
1.4.2	Purchase Inventory	I	A	C	I	I	R	I	
1.4.3	Stock Location	I	A	I	C	C	I	R	

Figure 4.6 Responsibility, accountability, consultative, and informative (RACI) matrix

The responsibility assignment matrix, in its simplicity, is a powerful organizational tool the project manager uses to manage human resources throughout the project life cycle and to effectively communicate responsibilities to stakeholders and the project team. The project manager may also find that if roles and responsibilities for project staff and stakeholders are not clearly defined and communicated, this can cause problems. For example, some individuals might think they have a higher level of responsibility and authority when, in fact, they do not, and other individuals who are supposed to have more responsibility are unaware of their role. The important element with a tool such as a responsibility assignment matrix is that it is an effective and efficient tool not only to communicate roles and responsibilities, but also to monitor and control individuals to ensure they are performing as expected.

Work Authorization

After an assessment is completed for all human resources involved in a project, it is vitally important that the project manager communicate clearly what authorization each category of responsibility will have. Authority is managed in different ways and is largely a function of how the organization is structured.

Authority by Organizational Structure

Organizations that have a PMO typically have an authorization structure established for project managers to use in assigning individual responsibilities to human resources. If the organization does not have a PMO, together the project manager and upper management generally make a determination as to what authority, if any, certain individuals on the project will have. Organizations typically manage work authority depending on how the organization is structured. Next, we look at three types of organizational structures and how authority is delegated from a project perspective.

Functional Organization

Work authorization is managed differently depending on the organizational structure. Organizations that have a functional organizational structure employing the traditional departments structure projects differently, and in most cases, work authorization and spending are the *responsibility of the functional managers*. The project manager coordinates human resources and work activities, procurements, and use of equipment and facilities through functional managers who have the authority to manage those resources and decisions. This type of structure can be an advantage for the project manager because the sole responsibility for work authorization is with functional managers who are trained and carry this responsibility on a regular basis. It can be a disadvantage for the project manager who has little authority over the departments, in this type of project environment, and may run into conflicts with scheduling work activity, allocation of human resources, critical procurements, and equipment and facility use.

Projectized Organization

Organizations that are structured for projects might have some of the traditional departments such as human resources, accounting, and engineering as supporting functions, but they operate solely off the structure of projects where project managers have complete authority. In this type of structure, individuals on a project are given work authorization responsibilities to manage specific areas within the project. This is an advantage for the project manager in having complete control over all individuals making decisions for scheduling, procurements, and use of facilities and equipment. The only disadvantage is the availability of individuals who have leadership or management experience, those who can be trusted to carry out work authorization correctly and appropriately.

Matrix Organization

Organizations that conduct business with a combination of projects and other business services or manufacturing may have project managers overseeing projects with a level of authority predetermined by upper management. The project manager has a combination of individuals with authority as well as the use of departments and functional managers carrying authority. The advantage of this model for the project manager is generally that there is a good selection of individuals skilled and experienced with the responsibility of authority, as well as a blend of key individuals and functional departments within the organization.

Authorization Definition

For each work package activity defined in the work breakdown structure, various components of that activity require some form of authorization. The project manager needs to consider, in the assessment of responsibilities, which individuals will be assigned various types of authority. The type of authority can be defined in general forms such as the following:

Scheduling—Authorization for scheduling falls under two general categories: human resources and facilities or equipment. Human resource scheduling is highly dependent on the organizational structure. In functional organizations, human resource scheduling is managed by functional department managers. In projectized organizations, typically the management and scheduling of human resources is done by project managers. In this environment, individuals move from project to project as they complete their tasks and are scheduled accordingly by the project managers. In a matrix organization, scheduling is typically a collaborative effort performed by functional managers and project managers.

Work activity—Authorization to perform work activity can fall within the tasks of any responsible person overseeing a work activity. Functional managers, project managers, and other organizational management and external subcontractors are examples of individuals empowered to authorize the beginning of a work activity.

Spending—Authority to utilize the organization's financial resources is typically dictated by the organization's executive management and/or the accounting department. Financial resources can be in the form of corporate lines of credit, monthly cash flow expenditures, individually issued credit cards, and petty cash within small departments. In most cases, executive management or accounting office staff establish who has authority to spend and determine a maximum dollar amount allowed based on responsibility level and/or individual.

Contract negotiation—Often, this area of authority can be an underestimated component of business within an organization. Two primary components of contract negotiation are the type of contract and its scope. Some contracts can be as simple as renting a small piece of equipment from a rental yard to be used on a work activity, whereas other contracts may be much larger in scope, requiring a great deal of background knowledge on what is being negotiated and the large sums of finance in play. It is usually a best practice for the project manager to solicit the advice of those experienced and skilled in contract negotiation for a particular element of the project to avoid contractual agreements that can put the project or the organization at financial risk. The project manager should be aware of all contracts on a project and control all contract negotiations during the project life cycle so that only the appropriate personnel are involved and can assist in mutually beneficial contract agreements.

Change control—All projects undergo requirements for change to some degree, and the project manager is responsible for ensuring there is a change control process in place. When an organization has a change control process, an individual or team identified in the change control process must correctly analyze all information and authorize changes accordingly. The project manager should be careful not to step out of the change control process and authorize changes solely on his discretion because doing so can impact the project schedule, budget, and quality of the project deliverable.

Risk contingency—Projects occasionally encounter risk events that have been identified within the project risk management system. In that case, contingency efforts were identified and have to be authorized. Because risk contingencies can mean the allocation of financial resources for the purchase of items, a new direction for a work activity, or the possible rescheduling of human resources, authorization for these types of actions should be issued only to individuals who have had a detailed briefing from the project manager on specific risk events and corresponding contingency plans. The authority for risk contingency has typically been planned in advance by the project manager, and individuals are named within the risk management plan as having the authority to carry out such contingencies. This is another area that it is vitally important that the project manager control to avoid random individuals making on-the-fly decisions about what to do in the event an identified risk event should happen.

In the development of human resource responsibility assignment, It is vitally important that the appropriate authorizations are assigned to individuals skilled and experienced with the authority they are given. It is also important the project manager effectively communicate what levels of authority individuals have to stakeholders and the project team to avoid confusion and/or misrepresentation of authority.

Information gathering and defining details of work activities can consume a great deal of a project manager's time and any project staff assisting in this process. Sufficient time must be allocated during this phase of the project to ensure the information gathering and analysis of work activities is not compromised. The project manager will learn a great deal about the activities required in producing project deliverables, the responsibilities and assignments for human resources, and the overall structure of how the project will be designed. He should leave this phase of planning armed with details of each work activity; the sequential order of work activities; and an understanding of human resource roles, responsibilities, and assignments.

Review Questions

1. Explain why it is important that the project manager and project staff have detailed information on each work activity.
2. What is the general purpose in defining responsibilities of those affiliated with the project?
3. Are there any drawbacks to using a responsibility assignment matrix?
4. Explain why authority on a project has to be defined and communicated with the project stakeholders and project staff.

Applications Exercise for Chapters 4, 5, 6, 7, and 8

Klanton Bower Data Center: Case Study

The Klanton Bower Data Center is a commercial data center specializing in small business IT services such as mass memory storage, email and enterprise systems servers, and website services. The data center, which has been in business for eight years, was started by the current owner, Klanton Bower, Ph.D., a data system specialist and research scientist. The data center also has four senior managers and several technicians and support staff who manage all the equipment and IT services infrastructure.

Dr. Bower and the senior management have been researching next-generation equipment for a possible expansion of their services to meet current market demands and to prepare for future next-generation requirements of large enterprise software platforms.

As Dr. Bower and senior management review the capability of current infrastructure equipment and what is required to upgrade and expand to meet future demands, this is considered a large-scale expansion of their current operation and is viewed as a special project. Dr. Bower has asked senior management to outline what equipment would be needed, provide estimated delivery dates of equipment, and create an outline of the events required to accomplish this expansion project.

Senior management understand the importance of purchasing the correct equipment and, from a project standpoint, correctly managing installation and testing as well as the go-live to have as little impact on the data center operations as possible and have a seamless crossover so that current customers will not be affected. In the initial outlay of logistics, it has been determined that one piece of critical server equipment has a lead time twice as long as several of the other pieces of equipment, and that may pose a challenge in the installation and testing of all the equipment. Fortunately, this piece of equipment can be installed at a later date and be integrated and tested at that time. The second challenge is clearing enough space in the current data center for installation of the entire expansion infrastructure.

The primary components of this project include finalizing research and the equipment list; procuring all equipment and establishing delivery schedules; organizing and clearing the data center floor for the expansion of equipment; and sequencing all activities required for the installation, testing, verification, and go live cross-over to implement the new equipment.

Case Study Exercise for Chapter 4

1. Identify work activities.
2. Collect critical information for each activity.
3. Define responsibilities.

5

Activity Sequencing

Introduction

At this point in the development of a project, the project manager should have broken down project deliverables into their smallest components called work activities, gathered information for each work activity, and arranged the activities in a work breakdown structure (WBS). Although it would seem as though the project manager could authorize the beginning of the project and the commencement of the first work activity, a critical step still remains in understanding the relationship that work activities have with each other and how these relationships influence the sequencing of work activities.

Sequencing work activities is a critical component in both scheduling project activities and managing the resources required for these activities. *Sequencing* is the process by which information gathered on each individual activity is analyzed to see whether there are relationships between activities that would suggest the proper placement of activities within the project schedule.

For example, in the construction of a foundation for a house, the cement cannot be poured before all the elements within the foundation such as trenches, plumbing, the electrical system, rebar, forms, and any other components have been completed within the boundaries of the foundation. Based on the relationship of the interior components of the foundation, it might be determined that trenches have to be dug first; plumbing, electrical, and building forms can be performed simultaneously; but all activities within the boundaries of the foundation have to be completed prior to pouring cement. Based on the requirements of these activities and the relationships of these activities to each other, conclusions can then be made as to the sequencing of activities in relation to cement being poured and the overall sequencing of foundation activities.

The project manager needs to understand other relationships such as activities that can begin only if a prior activity has been completed, while other activities can run simultaneously and not be dependent on one or the other being completed. Some activities might be required in the overall objective of the project but not be components of a particular project deliverable and how periphery items such as these are scheduled within the master project plan. It is important that the project manager understand the concept of activity sequencing because this is critical in scheduling and also project activity control to correctly produce project deliverables and accomplish the overall project objective. This chapter explores concepts and tools to understand project activity relationships and methods for the proper sequencing of activities in scheduling a project plan.

Information for Sequencing

The process of activity sequencing starts with the analysis of information gathered for each activity. To help understand what information is required to properly sequence activities, the project manager can look to the information gathered for each activity and documented within the work breakdown structure (WBS). The other valuable tool in determining relationships of work activities is the information gathered in the activity decomposition decision tree analysis. Between these two documents, the project manager can now take this information to subject matter experts, whether internal or external to the organization, to seek advice regarding the way activities relate to each other and gain insight as to the proper sequencing of activities based on these dependency relationships. Within the organization, the project manager might also find valuable archived information from past projects that had similar activities and be able to derive not only information of activity relationships, but also lessons learned as these activities were carried out.

As more information is gathered to understand the relationships between activities, this information should be documented in the same location as other activity information. This way, the project manager and project staff can reference a single location for all information concerning a project activity. Components of information covered within this chapter can be added to the *activity information checklist* outlined in Chapter 4, "Activity Definition." This allows individuals tasked with gathering activity information to be more efficient while collecting data and interviewing subject matter experts.

Activity Information Required

Much of the basic identification information required for proper sequencing of work activities can be found in the activity information locations within the WBS. Additional information that is required but not necessarily found in the initial collection of data for an activity might include

Relationship to other activities—Some activities may be dependent on prior activities being completed; these activities have to be completed before another activity can be started. Or the activities have no direct connection to activities for the project deliverable but are required in the overall project objective.

Constraints—These special conditions of an activity that have to be met can influence other activities, the critical timing of an activity in relation to other activities, and organizational or logistical complexities that might alter the characteristics of an activity.

Identified risks—Identified risk events concerning an activity can influence the start of an activity, its characteristics, or its overall duration.

This type of information plays a critical role in sequencing activities because it can give insight into the relationships that activities have with each other and the impact that characteristics of activities can have on neighboring activities and the overall project. It also plays a role in understanding the importance of each activity regarding whether or not it is critical in the path of all activities and whether special attention needs to be given to ensure it is completed properly and on schedule.

Diagramming Methods

A common way to represent and analyze project activities is through a visual representation of these activities connected together in the form of a network. Two processes are used to evaluate work activities for diagramming a network; they are called the *program evaluation and review technique* (PERT) and the *critical path method* (CPM). PERT was originally developed in the late 1950s in joint collaboration between the U.S. Navy, the Lockheed Missile Systems division, and the Booz Allen Hamilton consulting firm for use on the POLARIS missile program. PERT has historically been used in research and development; it emphasizes interdependencies between activities and allows the project manager to manage resources between activities. In most cases, due to its complexity and expense, PERT was used only by larger corporations. CPM was originally developed by DuPont Inc. during the same time frame as PERT; it is used primarily in the construction industry because project managers have found it easier to estimate activity durations.

Shortly after the development of PERT and CPM, the project management community began using a simpler technique called the *arrow diagramming method* (ADM) and its corresponding *activity-on-arrow* (AOA) diagramming technique or the *task-on arrow* (TOA) technique. More recently, this has been replaced with the *precedence diagramming method* (PDM) and its corresponding *activity-on-node* (AON) diagramming technique found in most project management software packages and covered in more detail later in this chapter.

Network Diagramming Terms

Throughout the world, industries and professional communities have developed terminology that is used uniquely to that industry or profession. The project management community has also developed terminology that is used in scheduling project activities; some of these terms are listed here to clarify definitions as applied to project scheduling:

Backward pass—The process of calculating late start and late finish dates for each work package activity connected on a path through the network diagram of a project. This process starts at the end of the network diagram and moves backward through all paths to the beginning of the network diagram.

Burst activity—A work package activity with two or more dependent activities (successors) in the network flowing away from it. Work cannot begin on the successor activities until all work on the burst activity has been completed.

Critical path—The path of activities connected through a network diagram that, combined, have the longest duration. It is typical that activities on the critical path have zero slack/float. It is possible to have more than one critical path.

Early finish date (EF)—The earliest point uncompleted work can finish on an individual work package activity.

Early start date (ES)—The earliest point uncompleted work can begin on an individual work package activity.

Event—An activity that has no start or completion and no resources assigned in no time duration.

Float/slack—The amount of time an activity's start can be delayed without affecting the overall project. Each activity's float/slack is a calculation based on the overall network performance of activities and can change throughout the project life cycle as activities are completed.

Forward pass—The process of calculating early start and early finish dates for each work package activity connected on a path through the network diagram of a project. This process starts at the beginning of the network and moves through all paths to the completion of each path.

Late finish date (LF)—The latest point uncompleted work can finish on an individual work package activity.

Late start date (LS)—The latest point uncompleted work can begin on an individual work package activity.

Merge activity—A work package activity with two or more dependent activities (predecessors) in the network flowing to it. The merge activity is not able to begin until all dependent activities reporting to it are completed.

Network diagram—A pictorial chart of project activities arranged and connected in sequential order based on logical relationships and activity requirements.

Node—The connection point of activities based on dependent relationships with other activities.

Parallel activities—Two work package activities that can be performed simultaneously having no predecessor or successor relationship dependencies between the two activities.

Path—A sequence of work package activities connected to form a flow of work activities based on dependent relationships.

Predecessors—Activities that must be completed before the next dependent activities connected in the network can begin.

Project deliverable—A compilation of completed work activities to form a final product identified by the customer as accomplishing a project objective.

Serial activities—A succession of single work package activities connected in one single line with equal predecessor and successor relationships.

Successors—Activities that cannot be started because they are dependent on prior activities connected in the network being completed.

Work package activity—The lowest-level activity in the breakdown of a project deliverable. The work package activity normally consists of an identifiable element of work that has to be completed and the expected time duration.

Defining Dependencies

The proper sequencing of work package activities can be accomplished correctly only if dependency relationships have been analyzed and determined for each activity. In the progression of completing work activities, completed work is combined together to form larger components of a project deliverable until it has been completed. Some elements of work can be completed with little or no dependency on other components within the project deliverable, whereas other elements are highly dependent on the completion of one work package activity before the next work package activity can begin. This process of defining dependent relationships between activities is very important in correctly sequencing activities. The first level of defining activity dependency is identifying the basic *predecessor* and *successor* activity relationship.

Predecessors and Successors

The *predecessor* relationship, as described previously, simply establishes one or more activities that need to be completed before the next activities connected in the network can begin. This definition is a simple yes or no qualification for activities that establish a definite sequence of work activity based on activity requirements and relationships to other corresponding activities. Figure 5.1 shows the predecessor and successor relationship as numbers identifying the previous dependent work activity. The predecessor relationship is also illustrated in the connection of activities shown in the network diagram in Figure 5.2.

Line Item	WBS Code	Work Activities	Duration	Predecessor
1	1.0	**Build Lawn Mower Project**	6.0 hr	
2	1.1	**Acquire Mower**	3.5 hr	
3	1.1.1	Define Requirements	0.5 hr	
4	1.1.2	Make Purchase	2.0 hr	3
5	1.1.3	Arrange Delivery	1.0 hr	4
6	1.2	**Mower Assembly**	2.3 hr	
7	1.2.1	Remove from Box	0.1 hr	5
8	1.2.2	Build Deck Assembly	0.5 hr	7
9	1.2.3	Build Handle Assembly	0.2 hr	7
10	1.2.4	Build Motor Assembly	0.5 hr	7
11	1.2.5	Final Assembly	1.0 hr	8,9,10
12	1.3	**Test**	0.2 hr	
13	1.3.1	Install Oil and Fuel	0.1 hr	11
14	1.3.2	Test Verification	0.1 hr	13

Figure 5.1 Predecessor and successor relationship in the WBS

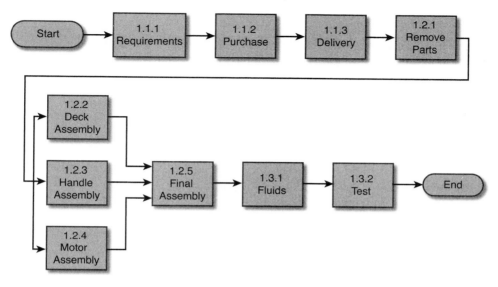

Figure 5.2 Predecessor and successor relationship for networks

Successor relationships are similar to predecessor relationships in that these are one or more activities that cannot start until a prior activity connected in the network has been completed. This also carries a simple yes or no qualification based on work activity requirements and relationships to corresponding activities. It is always tempting for the project manager to authorize the partial start of a successor activity when the prior activity has not been completed to keep project activities moving. However, this discretionary call by the project manager steps out of the boundaries established for work activity quality control and runs the risk of causing more problems than it solves. It is always best for the project manager to exercise caution and follow the predetermined dependency relationship each activity has with other activities in the network.

In the analysis of each work package activity to determine predecessor and successor relationships, the project manager or project staff may make decisions as to why a predecessor or successor was chosen. In this case, the relationship might not be a hard and fast requirement, but due to other reasons for creating the relationship. There are three reasons that dependency relationships are created for project activities—mandatory, discretionary, and external—as described next.

Mandatory

Mandatory dependencies are hard requirements of the work package activity and cannot be reversed. For example, all activities taking place within the boundaries of a foundation must be completed before cement can be poured to complete the foundation activity.

Discretionary

Discretionary dependency relationships can be established as a matter of discretion, deciding which activities should go before or after other activities. For example, the project manager may determine it is best to order all materials for building a house before the foundation is completed, or may find it best to order only some of the materials before the foundation is completed and the remaining materials after the foundation has been completed.

External

The external type of dependency involves relationships concerning project activities and external influences outside the project activity network that can influence an activity on the project. For example, in building a new house, the wiring for the phone system cannot be fully tested until the utility company successfully makes the connection to the city phone system.

Precedence Diagramming Method (PDM)

Over the years, the project management community has developed and used various types of diagramming methods in large and small corporations and various types of projects to effectively document and manage project activities, resulting in both success and failure. It is this same project management community that has continued to analyze diagramming methods and refine and improve these methods until they are successful across all sizes of corporations and project types. The precedence diagramming method (PDM) has surfaced as one of the most successful methods to understand and implement; it also is one of the more commonly used methods in project management today. This diagramming method is also easily incorporated in most enterprise software management systems that include project management functionality as well as smaller independent software packages that individuals can purchase, install, understand, and implement fairly easily.

The precedence diagramming method relies on two primary components of philosophy: the interdependencies of project activities and the various connections of project activities based on these dependencies. The primary tool used in graphically illustrating a PDM network of project activities is the activity-on-node (AON) method. AON relies heavily on correctly identifying the relationships and dependencies of all activities included on the network diagram.

Activity Relationships

To help understand the concept of activity dependency, we first review the two primary dependency types and the four types of relationships that activities might have based on these dependencies. The *predecessor* is an activity that has to be completed before the next logical activity in the network can be started. The *successor* is an activity that comes after and is dependent on a prior activity being completed. Based on these two types of dependencies, four activity relationships can be derived (see Figure 5.3):

Finish-to-Start (FS)—A relationship wherein a successor activity cannot start until a predecessor activity has finished

Start-to-Start (SS)—A relationship wherein a successor activity cannot start until a predecessor activity has started

Finish-to-Finish (FF)—A relationship wherein a successor activity cannot finish until a predecessor activity has finished

Start-to-Finish (SF)—A relationship wherein a successor activity cannot finish until a predecessor activity has started

Relationship Types	Description	Illustration
Finish-to-Start (FS)	Activity B cannot start until activity A finishes	 Activity A → Activity B
Start-to-Start (SS)	Activity B cannot start until activity A starts	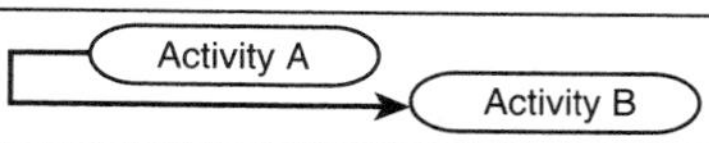 Activity A / Activity B
Finish-to-Finish (FF)	Activity B cannot finish until activity A finishes	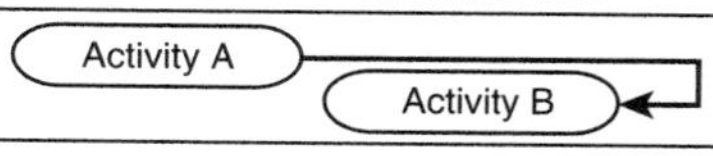 Activity A / Activity B
Start-to-Finish (SF)	Activity B cannot finish until activity A starts	 Activity A / Activity B

Figure 5.3 Four activity relationships

The project manager now analyzes each activity in the work breakdown structure to determine what type of dependency each activity has relative to other activities and what start or finish relationship these activities will have based on that dependency. To help organize this information, the project manager can use a simple tool called the *activity dependency matrix,* as shown in Figure 5.4.

Activity	Predecessors	Duration
A	None	4
B	A	5
C	B	3
D	B	2
E	B	3
F	C,D,E	4

Figure 5.4 Activity dependency matrix

Although the figure shows a simple form of this matrix, the project manager can design the activity dependency matrix to include other information that might help in developing the network diagram.

Activity-on-Node (AON)

The network diagram is an excellent tool to understand the details of activity relationships within the project and understand other critical information such as overall duration of the project, critical path relationships, and any float/slack that might be available for activities. Before the project manager can begin developing a network diagram, it is important that critical information of each activity has been acquired such as an activity label and that the time duration of each activity and dependency relationships (predecessor, successor) have been identified. Using the information included in the activity dependency matrix, the project manager can now begin to formulate the structure of a network diagram using the activity-on-node (AON) method. Figure 5.5 illustrates how an AON network diagram is constructed, how activities can be connected using arrows to define basic relationships of activities, and how pathways can be developed through the network (based on information in Figure 5.4).

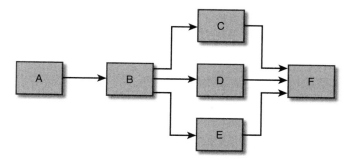

Figure 5.5 Basic network diagram

Before construction of the network diagram can begin, there are some basic rules that can be helpful in not only understanding how the network diagram is constructed but also how the diagram actually works:

- A minimum amount of information needs to be gathered for each activity and arranged within a tool such as the activity dependency matrix.
- A labeling convention needs to be established to display information for each activity on the network.
- The network diagram typically flows starting at the left and moving toward the right.
- Connections between activities use arrows to identify predecessor and successor relationships.
- Activities can be used only once within a diagram. If the same type of activity is performed multiple times within a project life cycle, it is best to label each one uniquely so they can be identified within the network diagram.
- After the network diagram is complete and the project has begun, information about each activity needs to be updated and may result in changes to activity relationships, durations, activity float/slack, and the critical path, which can result in updates to the overall duration of the project.

Activity Labeling

Using the activity-on-node (AON) diagramming method, the project manager first needs to create the *node*, which is the identifying label on the network documenting information about a work package activity. Reviewing the information listed on the activity dependency matrix indicates which nodes have to be created and where arrows will be placed to connect nodes based on their dependency relationships. Project network diagrams can have several work package activities displayed as nodes; consequently, it is best to maintain a minimal amount of information on a node in the interest of keeping the nodes small enough to help manage the overall size and complexity of the network diagram. Figure 5.6 illustrates a work package activity node and the corresponding required information locations within the node.

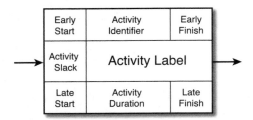

Figure 5.6 AON node information

After the project manager understands why certain components of information are required within the node, nodes can take on different shapes, and information can be arranged slightly differently based on the project manager's individual style of designing and using the node within the network diagram. Activity nodes used in software programs such as Microsoft Project might be shaped differently but have the same type of information and are used in the same way within the network diagram.

Activity Path Definition

As work package activity nodes are being created and linked together based on their dependency relationships with other activities—a network diagram—begins to take form. As more nodes are added to the network moving from left to right, pathways start to develop, indicating the sequence of activities required to complete elements of the project deliverable. In AON network diagrams, there are four variations of paths, depending on dependency relationships: series, parallel, burst, and merge.

> **Serial activities**—These work package activities have been identified to be in one single succession of events and have a single predecessor and successor relationship. These activities (A, B, C, D) typically have no other dependency requirements, and no other activities are attached to them, as shown in Figure 5.7.

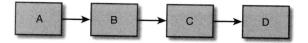

Figure 5.7 Serial activities

Parallel activities—These work package activities have been identified to have no distinguishable dependency between them and therefore have no requirement to be connected in the network in succession. These activities (B and C only) can be performed simultaneously, as shown in Figure 5.8.

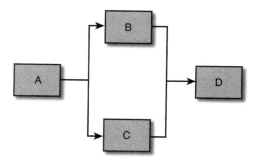

Figure 5.8 Parallel activity

Burst activity—This work package activity has been identified to have two or more immediate successor activities flowing away from it. All successor activities (B, C, D) can start only when that burst activity (A) has been completed, as shown in Figure 5.9.

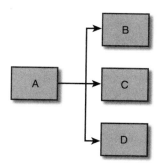

Figure 5.9 Burst activity

Merge activity—This work package activity has been identified to have two or more immediate predecessor activities flowing to it. All predecessor activities (A, B, C) have to be completed before the merge activity (D) can begin, as shown in Figure 5.10.

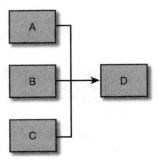

Figure 5.10 Merge activity

With many projects, most of these activity dependencies and relationships exist, so it is important that the identification and labeling of each activity are correct. The reason is that the labeling can have a drastic impact in how the succession of work activities is carried out and in the overall planning and duration of the project. As more nodes are added onto the network diagram, various pathways that begin to emerge in the network diagram can become more complex, having multiple pathways based on activity dependencies and relationships. Figure 5.11 illustrates a relatively simple network diagram for a project but shows all four types of relationships and corresponding paths that can be produced.

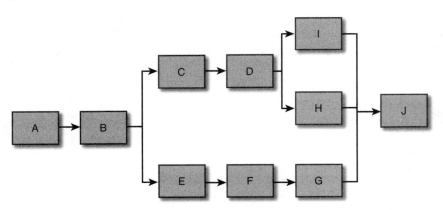

Figure 5.11 Network diagram using serial, parallel, burst, and merge activities

Determine Critical Path

After the network diagram is complete and all connections between activity nodes are confirmed, an analysis can be performed to determine pieces of information about each activity in the overall project. The project manager and other management staff with responsibilities over work package activities want to know information such as how critical start and finish times are, whether there is flexibility with durations of each work activity, and whether there will be scheduling constraints based on one activity's dependency on one or more other activities. In many cases, this type of information can be difficult to ascertain at the beginning of a project during the information-gathering process because dependencies and activity relationships are sometimes difficult to define. It is generally an enlightening moment, at the completion of a network diagram, to see how many interdependencies and constraints there are and how important it will be to manage activities within the project.

Although a large number of individual work package activities can be required to complete a project deliverable, another component of creating the network diagram is revealing the chain of activities that might be most critical in maintaining the project schedule and therefore those activities that the project manager needs to pay close attention to. As work activities are analyzed throughout the network diagram, the project manager will ultimately want to know which path represents the longest path, which will establish the overall duration of the project; that is called the *critical path*.

The term *critical path* is used commonly in project management, and although it does mean the longest path through the network diagram (project schedule), it should be isolated to that definition alone. The critical path usually represents activities that have no leeway in duration or have zero float/slack; consequently, if activities on the critical path fall behind schedule, they will affect the overall duration of the project.

It is also important to note that although the critical path represents the longest path through the network diagram, it does not necessarily represent other critical elements of project activities and the overall project.

The critical path does not represent

- The longest duration activities
- All activities critical to the project
- All activities with the highest cost
- Activities with the highest risk potential

To better understand what the critical path is and how it is used in managing the project, we look at the analysis of how the critical path is derived and what information about each activity can be gathered from an analysis of the network diagram. To begin this process, let us start at the far left of the network diagram and move to the right through each path and determine how many paths there are through the network and the duration of each path, as shown in Figure 5.12.

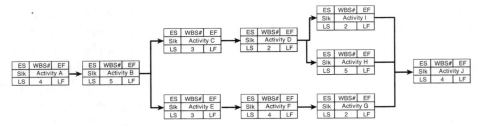

Figure 5.12 Identify all paths and the critical path

Path #1: A, B, C, D, I, J	Total Path Duration	4 + 5 + 3 + 2 + 2 + 4 = 20
Path #2: A, B, C, D, H, J	Total Path Duration	4 + 5 + 3 + 2 + 5 + 4 = 23
Path #3: A, B, E, F, G, J	Total Path Duration	4 + 5 + 3 + 4 + 2 + 4 = 22

The longest path is #2 (23), so this is the critical path.

In identifying all paths through the network, one might determine that there is more than one critical path. Upon the discovery of this phenomenon, it is best to perform the path calculations a second time to confirm duration values and the existence of a second critical path. It is theoretically possible to have more than one critical path, as shown in Figure 5.13, but this is a rare occurrence for most projects. The project manager must be aware that if multiple critical paths have been identified, this is only at the onset of the project, and as activities are completed and updates for activity durations have been made, path characteristics can change over the course of the project life cycle. A critical path that was identified at the beginning of the project may cease, and a new critical path may emerge based on new information of work activities. In the case in Figure 5.13, activity G is taking longer and is now three days, which has created two multiple paths.

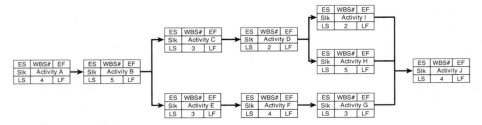

Figure 5.13 Multiple critical paths

Path #1: A, B, C, D, I, J   Total Path Duration   4 + 5 + 3 + 2 + 2 + 4 = 20
Path #2: A, B, C, D, H, J   Total Path Duration   4 + 5 + 3 + 2 + 5 + 4 = 23
Path #3: A, B, E, F, G, J   Total Path Duration   4 + 5 + 3 + 4 + 3 + 4 = 23

There are now two critical paths, #2 and #3, at 23 days.

Activity Analysis

After the initial analysis is complete on the network diagram, all paths are identified, and the critical path is determined, more information can be gathered about each activity that will be important in not only managing but also understanding the flexibility that might be allowed on each activity. Gathering this information requires two processes: the *forward pass* and *backward pass*. These two processes yield information on each activity such as early start (ES), early finish (EF), late start (LS), late finish (LF), and available float/slack.

Forward pass—The following example illustrates how a forward pass process is conducted. Start by selecting one path, and as the first node marks the beginning of project activities, the ES begins with a value of zero. Adding the duration of that activity onto the ES produces the EF value:

ES + Duration = EF

Follow the arrows to the next successor activities (nodes) and insert the EF value from the prior activity as the new ES value for the successor activities. If a node has two or more predecessor arrows flowing to it, and those predecessors have different value EFs, always record the *highest* value EF as the new ES. Add the corresponding duration to those ES values to derive their EF value. Continue the forward pass process for each path in the network diagram, as shown in Figure 5.14.

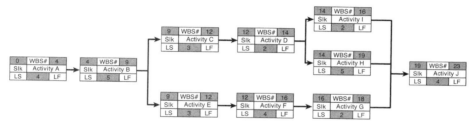

Figure 5.14 Forward pass process

Backward pass—The following example shows how the backward pass process is conducted. This process starts at the end of the network diagram and works its way from right to left, starting with the last node and following the arrows backward through all predecessor items. Record the end duration value for all immediate predecessor LF values. Subtract the duration of the activity from the LF value to derive the LS value for that activity:

LF − Duration = LS.

If two or more activities feed into one predecessor, the *smaller* LS value is transferred to the activity as the new LF value. Follow the arrows to the next predecessor activity (node) and perform the same calculations continuing backward through the network until the starting node is reached, as shown in Figure 5.15.

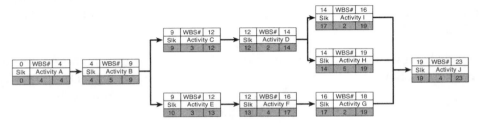

Figure 5.15 Backward pass process

Float/Slack Calculation

Upon completion of the forward and backward pass processes, the project manager can now analyze information for all activities as to the earliest possible start date and finish date as well as the latest possible start and finish date. Having this information for each activity allows the calculation of float/slack for the start and finish of each activity. The calculation for start float/slack is SLK = LS − ES, and for finish is SLK = LF − EF, also shown in Figure 5.16.

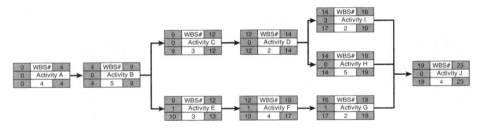

Figure 5.16 Float/slack calculations

After the float/slack calculations are made, this also is a way to confirm the critical path and that it has little or no slack. The project manager then needs to manage activities on this path to ensure the project stays on schedule. See Figure 5.17.

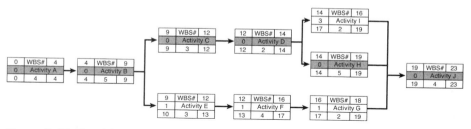

Figure 5.17 Float/slack confirms critical path

Activity information derived from the network diagram and analysis process can be invaluable for the project manager. It enables her to understand details of each work package activity and the relationships of activities and interdependence of these activities within the project. The project manager and those responsible for overseeing individual work package activities can use this information in several ways, including the following:

- Confirm all work package activities have been accounted for.
- Confirm the relationships and dependencies for each activity in relation to other activities.
- Confirm contractual start and stop time commitments are in the correct sequence in the project timeline.
- Understand the need for special attention to managing activities listed on the critical path.
- Understand the possible availability of float/slack on start and finish times for each activity.

The project manager and other management staff within the organization also can use information from the network diagram and activity interdependencies for other analyses, such as the following:

- Analysis of cash flow for the project
- Analysis of resource allocations
- Analysis of potential risk impacts on other activities and the ripple effect that risk can have throughout the project life cycle
- Recalculations of overall project duration and estimates at completion

The AON network diagramming method is a relatively simple tool and can be produced within most project management software programs such as Microsoft Project; it is typically derived from the initial WBS information. In many cases, the success and/or failure of a project can be traced back to the information-gathering process as well as the information analysis process concerning work package activities. Simple tools such as the network diagramming method have been developed so the project manager can use work activity information wisely and constructively to understand project dynamics and areas available for project control.

Review Questions

1. Explain what activity information is important for sequencing activities.
2. Briefly explain each diagramming method.
3. What is meant by *activity dependencies* and how do they relate to network diagramming?
4. Explain the relevance in the critical path.

Applications Exercise

Klanton Bower Data Center: Case Study (Chapter 4)

Apply the concepts described in this chapter to the case study:

1. Define any activity information that would be useful in sequencing activities.
2. Determine a diagramming method.
3. Identify predecessor relationships, if any.
4. Create a network diagram of activities.

6

Resource Estimating

Introduction

Resources make up the backbone of a project. From human resources to facilities, cash flow and lines of credit to equipment, patents to proprietary and intellectual knowledge—all have to be selected carefully to ensure their value not only to the organization but also to each project. Resources are what the organization uses to structure a project to complete its objective. Managers are responsible for overseeing processes being performed within the operation, so how effectively they select and manage resources plays a critical role in project success.

Resources can vary greatly depending on the size of the organization, how it's structured, and the approach that executive management staff want to take in designing the organization. Some organizations, such as service companies, are very human resource oriented, whereas others, such as manufacturing and construction, might be more equipment or facility intensive. Functional and project managers have to allocate the right resources effectively and in a timely fashion to accomplish their objectives. This includes the selection and training of human resources and the purchase and allocation of capital equipment and facilities. Of these different types of resources, generally human resources present the greatest challenge in effectively managing skill sets and scheduling for project activities.

Human resources are part of every organization at some level because *people* create the organization, manage it, and perform tasks and processes associated with it. Some companies are more automated and require few human resources to perform tasks, but generally there is at least one human who started the organization. Human resources, in some ways, are similar to other resources such as equipment, machines, and software because they serve a particular purpose or perform a specific task.

However, they do have unique attributes that make selecting and managing them a little more interesting and challenging.

Human resources can be among the hardest types of resources to manage because there are many variables to consider. They do have certain characteristics that are more fundamental with any type of resource. All types of resources share some common attributes; they may be

- Qualified for the task
- Readily available
- Cost effective
- Reliable
- Permanent or temporary

Many types of resources share these attributes because they are primary to acquiring resources. Attributes such as availability, cost, and permanence or temporariness are easier to assess because they are more absolute forms of information. Attributes such as qualifications and reliability may not be as easy to ascertain because it can be difficult to quantify them for any type of resource if a specific resource has not been used previously. This assessment may require some research in refining the type of resource needed or gathering information from references having used that type of resource to better understand its use and reliability. These attributes are considered more fundamental and not subject to change without notice in most cases, given most types of resources.

In the case of reliability, equipment and machines may break down, but this failure can usually be associated with poor maintenance or just an unforeseen incident. Selecting resources that have acceptable performance and reliability should result in the outcome desired if they are maintained properly. One type of resource that doesn't always fall into this category—because any number of influences can change the performance and reliability—is the human resource. This type of resource has the potential to vary in several different areas as a result of thought processing and mindset that other types of resources do not have.

All resources have to be evaluated regarding their ability to perform the task required. Usually, this information is available when acquiring the resource and is fairly accurate, with little change or variation. A human resource's ability is more difficult to evaluate because the initial information can be rather subjective and may require further evaluation for specific abilities and performance. Given this more subjective nature of human resources, even with in-depth evaluations, the full knowledge of a resource's ability will not be realized until used in the organization.

Project managers have the arduous task of evaluating all resources used on a project and rely on intuition, past experience, other functional managers, and "lessons learned" documents to assist in the selection process. The selection process itself involves the evaluation of fundamental characteristics of all resources and actual resource requirements as defined in work package activity information. Those tasked with the evaluation and selection of resources used on a project must understand the different types of resources and develop processes for the evaluation and selection.

Types of Resources

Organizations are designed to accomplish their strategic objective and acquire resources accordingly. If the purpose of an organization is research and development, it staffs its organization with engineers, large allocations of laboratory space within its facility, and specialized equipment for testing. An accounting firm has a totally different structure within the facility, uses different equipment, and requires a different type of staff. Construction companies also have a heavy allocation of equipment, staff certified to use that equipment, and large spaces to store equipment and materials.

So it is the organization's type of business that largely dictates what types of resources are required. Projects operate in much the same manner because each project has an objective, and the very nature of that objective dictates the types of resources required to accomplish it. Information gathered on each work package activity should indicate what types of resources are required to accomplish each activity. It is also important that subject matter experts help define resources for each activity to clarify minor details regarding critical characteristics required for specific resources.

Define Project Resources

Because an organization may have several types of resources at its disposal, the definition of work package activities reveals very specific requirements for certain resources. All organizations that conduct projects have six primary classifications of resources:

Human resources—These resources are individuals that the organization employs to perform tasks for the operation and/or projects. In the initial evaluation of potential human resources, it is very important to look at both depth and breadth of skills because this is the highest value the resource can bring. With projects, human resources perform one specific task or several tasks utilizing a broad base of skills, and they are allocated based on availability. Having resources with a wide range of skills allows management to better plan resource allocation within departments and for projects. This puts more emphasis on the evaluation process, when hiring, to better understand potential resources. Because most managers do understand this importance, the goal here is to connect the value of the skill set to resource utilization.

Financial resources—These resources are the means to purchase items required for project work activities. Financial resources can be in the form of large corporate lines of credit, credit available through suppliers, and credit cards that have been issued to individuals within the organization. Financial resources can also be in the form of corporate checks and petty cash funds to manage smaller, more immediate purchases. The accounting department in most organizations determines what types of financial resources are authorized for project procurements and expenditures.

Capital equipment resources—The organization has acquired these resources to carry out the activities of business, which are to accomplish the strategic objectives of the organization. Capital equipment can be in the form of office equipment and furniture; manufacturing machines; computer systems; specialized laboratory equipment; heavy equipment (for the construction and transportation industries), forklifts, trucks, and trailers; and warehouse shelves and pallet racking. The organization can either purchase or rent this equipment for short-term usage such as a project work activity.

Materials resources—These resources are items that the organization purchases for use in the manufacturing or distribution of project deliverables to customers, or consumables used for service-related deliverables.

Facilities resources—These resources are any structures that the organization has acquired to house all resources required to carry out business activities of the organization, which can include project work activities. They can be in the form of residences or commercial or industrial structures, such as office buildings, factories, and warehouse structures. Facilities can be purchased outright or can be leased for a period of time. One key element is the allocation of facilities space for projects that need to be shared with other functional operations. This is sometimes a constraining item and can create conflict between project managers and functional managers.

Information technology—This type of resource is unique in that it can be quite different depending on the organization. Through special research and development or customer requirements, some organizations have developed technology that is unique and has been patented for their use. This type of information puts the organization at an advantage when negotiating project deliverables because this information is typically unique to this organization. Another form of information technology is intellectual knowledge of certain individuals within the organization. This knowledge is usually found within the senior or executive management structure, engineering staff, or any other individuals who have been exposed to specialized information unique and proprietary to that organization.

Direct and Indirect Project Resources

As the project manager is evaluating each work activity for the required resources, it becomes evident that certain resources are directly related to work activity requirements and other resources that will be utilized do not necessarily have a direct connection to a work activity. Within some organizations, project managers have to account for all the resources allocated to a project to maintain cost and scheduling accountability. To perform this task, the project manager needs to ensure that all resources fall under one of two general categories:

Direct project resources—These resources are acquired and allocated to specific work activities based on those activities' requirements. The success and completion of work activities depend on direct project resources being allocated. Examples are certain human resources who are identified, based on skill set, to complete certain activities and particular pieces of capital equipment that have to be used to complete an activity.

Indirect project resources—These resources might be used during the course of project work activities but are not allocated to a particular work package activity. Examples include a facility housing all project work activities; the accounting department that might facilitate financial resources but is not directly involved in a work activity; and trucks and trailers that the organization owns that transport several project deliverables to customers and are not allocated to one work activity.

Contracted Resources

If an organization conducts a project and does not have a resource that has been identified in a work package activity, then that resource might have to be acquired through an externally contracted procurement. This procurement might be in the form of contracting human resources with specialized skill sets and/or knowledge required for a work activity. Or it might be in the form of general laborers hired to decrease the workload of project staff to improve the project schedule. Other external resources that might have to be procured through contract include the lease of additional facilities space, the lease of specialized equipment, or the rental of heavy equipment required for a particular work activity. In these cases, the organization might not have this equipment due to the infrequent nature of this particular work activity and therefore determines it is more feasible to simply contract specialized resources only when required by project activities.

Resource Constraints

The success of an organization is typically gauged on how the founders of the organization structured operation and how current executives manage activities to accomplish the organization's objectives. In the best interest of structuring an organization correctly and efficiently, executives discover, on a daily basis, the problems, pitfalls, and constraints that do not allow the organization to be as efficient and successful as they would like. Project managers run into the same problems on projects where constraints impose limitations in how projects can be designed and in the daily managing of project activities.

Project managers first discover constraints early in the project life cycle when details of a project deliverable are first being analyzed for feasibility. In most cases, the project manager and/or initial stakeholders begin to understand the limitations that the organization may have in the ability to carry out a potential project objective. Therefore, he must be careful in authorizing a project so as not to put the organization or the customer at risk of failure. The first sign of problems often occurs when those tasked with making the decision to authorize a project do not fully understand the impact a project can have on an organization or the ability of the organization to carry out a project, and they do not take identified limitations and constraints seriously. One key factor in signing off the commissioning of a project and the ultimate success of that project is not only understanding how the organization would be able to complete a project objective, but also what limitations the organization has in not being able to complete a project objective.

When one is analyzing constraints and limitations that an organization might have in successfully completing a project objective, constraints are associated with three primary categories: *organizational*, *project*, and *resource constraints*.

Organizational constraints—These influences are related to the overall structure and/or management of the organization. This area of influence is the largest in scope, encompassing aspects of the organizational structure, owners or founders of the organization, and general philosophy within executive management. Specific constraints from an organizational standpoint might include the following:

- Executive and mid-level management's lack of vision within a particular market can make it difficult to correctly select project opportunities that will represent value to a market. This can be in the form of missing market opportunities or misinterpreting market demands.
- Executive and mid-level management's lack of support for a particular project can have a huge impact on the success of a project because most of the project authorization is evaluated by this level of management. If this level of management does not agree with decisions made in how to carry out project activities, this can create conflicts and delays affecting the overall success of the project.
- The project objective is not within the scope of the organization's strategic objective. All projects have to be evaluated to ensure they align within the organization's strategic objective and therefore bring value to the organization when completed.
- The company is not organized enough to effectively carry out a project. Not all companies are structured such that there is an adequate level of general organization required to develop a structured project and carry out a project management plan accurately and effectively. In this scenario, the project would be doomed for failure at the beginning due to a lack of organization, planning, and structure.
- Extreme complexities of the organization might make it difficult to develop and manage a project. Some organizations have extremely complex operational components that could present difficult challenges in structuring and implementing a project. Organizations with highly proprietary information or government agencies dealing with classified information may encounter limitations in having complex relationships with other internal and/or external resources.

Project constraints—These influences are associated with elements of project management as they relate to a particular project. Although the organization may be well equipped and structured to carry out projects effectively and efficiently, individual projects have inherent limitations and constraints that influence the effectiveness of carrying out a specific project objective. These types of constraints are not necessarily influenced by the organization and are not resource-related constraints either, but more in the development of the project, management of the project, and an area typical of all projects called the *triple constraint*. Areas of influence on a particular project can include the following:

- The project objective does not align with organizational goals and objectives.
- The project scope is too large for the organization's size and capabilities.
- The project management plan is poorly designed.
- The triple constraint is typical on all projects because it represents three common elements of each project that have to be carefully managed within the project itself: schedule/time, budget/cost, and deliverable/quality (see Figure 6.1). The project deliverable is made up of activities that will be completed at a certain quality level, which has an associated cost and a predetermined time frame for completion. These three constraints are tied together because any change in one can affect one or both of the other elements, creating a constraint. If all three elements are operating within the parameters of the project design, they do not pose a constraint.

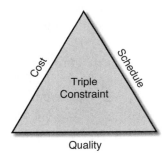

Figure 6.1 Triple constraint

Resource constraints—Influences from resource constraints can be classified as all other aspects of organizational operations that are directly or indirectly involved in project activities. These constraints can also represent aspects of the project that are internal or external to the organization. Although each type of resource can present certain specific characteristics of influence to a project, all resources present two main constraints to a project: *capability* and *availability*.

- **Capability**—All resources—whether human, capital equipment, facilities, or financial—have both abilities and limitations that can present opportunities as well as constraints to a project. All resources selected for use on a project are evaluated for the capability of performing a function that will have a positive contribution to project activities. Human resources, for example, are evaluated on abilities based on skill sets, background, and experience they have acquired that will allow them to complete specific work activities. Capital equipment is evaluated on abilities based on the design, functionality, and reliability that the manufacturer has built into that piece of equipment so that it also can complete specific work activities. Other resources, such as finances that can be in the form of lines of credit, are also evaluated for their ability to perform as required for specific procurement activities. It is important all resources are evaluated and selected based on the ability to perform specific tasks because this is paramount in efficiently and effectively accomplishing the project objective.

- **Availability**—Second to the capabilities of all resources required for a project is the availability of these resources. The project manager and other project staff, along with functional managers within an organization, can identify all resources required for the project. If, however, those resources are not available or are available at the wrong times throughout the project, this can present constraints to project schedule, cost, and in some cases, quality of the deliverable. Project managers spend a great deal of their day-to-day work managing the schedules of all resources associated with project activities.

In some cases, project managers working in collaboration with functional managers might have to negotiate resources based on schedule and availability. In the case of human resources and possibly capital equipment, this negotiation may result in settling for a resource that has a lesser capability but is available at the right time versus a resource with higher capability that is not available. Project managers may find that they do not live in a vacuum and have to share resources with other projects or other functional managers within the organization, and this need to share may present scheduling conflicts with work activities on their project.

At the beginning of a project, the project manager must understand the importance of communicating the project schedule to all applicable staff and confirming the availability of all resources identified to help minimize scheduling conflicts, cost overruns due to unplanned acquisitions of resources, and poor quality on work activities as a result of having to compromise by acquiring available resources.

Resource Requirements

Organizations are structured using resources that are obtained based on matching resource capability with demand requirements. If an organization is going to manufacture a product, it needs to have design engineers on staff to transition conceptual ideas to working prototypes. The organization also needs manufacturing and process engineers who can design production environments using facilities and equipment so that prototypes can be manufactured at higher volumes and efficiencies, thus allowing the organization to profit from the sale of the product.

Organizations require management structures to obtain resources to manage and carry out day-to-day operations. Each area in an organization, such as administration, accounting, human resources, procurements, warehouse, and shipping and receiving, requires facilities, capital equipment, and human resources to carry out activities. An organization that is service related has specialized personnel trained and skilled in the services that the organization provides; it also must obtain equipment, materials, and vehicles required to carry out these activities.

So as organizations require various types of resources to be successful, project work activities also require specific capabilities of those resources to effectively and efficiently complete project deliverables. It is clear that each individual resource must have a required capability and be obtained based on that resource's ability to perform required tasks. As previously stated, this requirement not only applies to human resources, but also is required of facilities and equipment, software, financial resources, and any other resources required to carry out project work activities. Challenges for most organizations with regard to resources typically fall within two categories: (1) the ability to acquire adequate resources cost effectively and, (2) internal to the organization, the effective allocation of resources. So there are general resource requirements for functional operations within the organization called *operations resource requirements* and specific resource requirements for project activities called *project resource requirements*. This chapter focuses on the latter.

Project Resource Requirements

Project deliverables have to be broken down into their smallest components to identify what specific work activities are required and what resources are needed to carry out work activities; these are called *project resource requirements*. Based on the work requirements to be performed within each individual activity, resources are selected as to their capability to fulfill the required work for that activity. It is important that the project manager and other project staff who assist in evaluating resources to fulfill activity requirements pay close attention to the details of resource requirements because this factor can play an important role in the selection of resources.

For example, if engineering work is required, an easier and lower-level type of work can be accomplished by less experienced engineers, whereas a more sophisticated and higher-level type of work requires an engineer who has a more specialized background and skill set. At first glance, the requirement simply should be acquiring an engineer. However, this task can get more complicated because higher-level engineers may not be available through internal resources in the time frame that they are required. Likewise, contracting external resources that do not have a problem with availability would cost the project a lot more. This is a typical dilemma with most resources required on projects because the constraints of using internal resources fall on scheduling and resource allocation, while the trade-off is the higher cost of acquiring external resources that will be available when needed. Part of the answer to this problem is understanding what critical capabilities the resource requires that can make or break whether the resource is available internally or whether obtaining the resource externally at a higher price is the only option.

Organizational Resource Management

As resources that are capable of completing work activities are being identified, the availability of these resources needs to be considered not only at the project level, but also at the organizational level. As previously stated, the availability of resources internal to the organization can largely be affected by how the organization is structured. Functional organizations having resources assigned to functional departments require the project manager to negotiate scheduling arrangements that can impose constraints not only to the project, but also to the functional department. Projectized organizations typically allocate resources to projects, and project managers and/or program directors are responsible for scheduling resources across projects. In either case, projects that require resources internal to the organization can be constrained with scheduling challenges and critical resource allocations.

Some organizations designate certain types of resources as "critical." They permanently assign these resources to an activity within the organization and do not make them available for use on projects. In some cases, organizations intentionally operate with fewer resources and procure external resources as needed to alleviate scheduling conflicts. Functional organizations are more typical in that they require projects to obtain external resources so as not to create scheduling conflicts internally. Management staff within the organization usually determine whether functional departments have priority with corporate resources. They also determine how and when these resources are available for projects and projectized organizations, where project managers hold a much higher level of authority and negotiate the allocation of resources between projects. When organizations fall short of having required resources, or negotiations result in resources not being available for projects, project managers must resort to external contracted resources to fulfill project activity requirements.

Resource-Estimating Methods

During the information-gathering process for work activities, resources were identified, and certain information pertaining to the requirements of each resource was also identified. It is important that specific characteristics of each resource required for an activity are documented because this plays an important role in how resources are acquired and allocated to the project. As previously stated, the organization may have certain resources available internally, and the identification and allocation of these resources are, in most cases, based on the negotiation of availability and scheduling. If the organization does not have a particular resource or it is unavailable, that resource needs to be acquired externally.

The process of estimating resources involves developing a method to define the utilization of available resources for a particular work activity within a defined time frame. This method not only includes resource availability, but can also include other attributes, such as the experience and skill set of human resources, facilities and equipment, the specialized options of functionalities, and general reliability. Defining the utilization of resources is an important component of developing the overall project management plan because the project manager needs to have commitments from the source of each resource to ensure they have been scheduled and will be available when required.

Because the project manager is primarily concerned with scheduling the allocation and utilization of resources for his particular project, depending on the size of the organization, the allocation of resources can be an easy endeavor or a very complex organizational process that requires higher levels of evaluation and approval. As you know, an organization might have in place three levels of project management structure, and each level has its own characteristics of resource utilization:

> **Portfolio resource estimating**—The portfolio level of project management structure within an organization is typically a very high level. Depending on the size and structure of the organization, this can mean an entire division of a company with several hundreds or thousands of employees. In some cases,

depending on the size of the portfolio and all the resources, human or otherwise, this level may include the portfolio director to be included on certain critical resource allocations and scheduling. If the portfolio is of such a size that this level of management is typically not involved in decisions regarding resource utilization, those decisions fall on lower levels such as program directors, project managers, and functional managers. In some cases, having specialized resources within a portfolio is an advantage because they are obtained for a specific function and are used within the organization only on projects within that portfolio and can be scheduled accordingly. The disadvantage of resource allocation within a portfolio is that the larger numbers of program directors, project managers, functional managers, and portfolio directors that can make decisions for resource allocation, the more complex it is to create resource constraints within the programs and projects of the portfolio.

Program resource estimating—The program level of project management is more a mid-level organizational structure and may have some authority in deciding resource utilization, depending on the type of organization. Unlike portfolios that can have several programs, projects, and other work activities that may be unrelated, programs are unique because the projects and work activities within programs are related. Typically, the organization obtains resources for specific activities required within projects of a particular program. This can be a great advantage for the program director and project managers because resource allocation within a program can stay within the negotiation between individual project managers and the program director, keeping the complexity of decision making at a minimum.

Project resource estimating—The project level within the organization is the lowest level of project management. The project manager has to develop a resource utilization plan and seek out approval and scheduling of all resources for each work activity. This can mean resources identified for a specific use on that project, which are easy to schedule, but also can mean the difficult negotiation of other resources that will be shared on other projects and/or used in other functional departments. This negotiation can result in constraints as to the availability of those resources for specific project activity time frames. If a project is not associated with a particular program or portfolio where program directors and portfolio directors can assist in the allocation of resources, projects are typically referred to as *standalone activities*, and the project manager has to negotiate resources independently for that project.

The project manager analyzes the utilization of several different types of resources used on work activities. In doing so, he requires certain methods, depending on the type of resource, the availability of resources, and how to gain approval for allocation and scheduling of all resources required. Several of the more commonly used methods in resource estimating are described next.

Delphi Method

A simple and commonly used method of resource estimating is called the *Delphi method*. The Delphi method was developed in 1969 by the RAND Corporation as a group consensus decision-making process. It incorporates the development of a group of subject matter experts (SMEs), managerial staff, and others internal or external to an organization who would have specific information of a project work activity and knowledge of corresponding resources that would be evaluated as the best solution for that activity. The key element in the success of the Delphi method is the dialogue between the panel of experts that results in the reconsideration and narrowing-down process of potential resources to the "most appropriate resource" based on the consensus (*expert opinion*) of the panel.

Determinate Estimating

Depending on the size and structure of an organization, resource allocation often falls under a predetermined association of resources and tasks that will always be carried out for activities given any project. These types of resources are easily identifiable within the organization because they serve only one purpose and are utilized for a specific task on a work activity. These resources typically are scheduled for allocation to activities throughout the year, and the availability for scheduling on new projects can easily be determined.

Determinate estimating is easier in organizations that are more project structured versus functionally structured. The reason is that the projects can be scheduled in a sequence where predetermined requirements for resources can be allocated to projects based on the sequential nature of scheduling all projects. Functionally structured organizations may have some resources with fixed schedules throughout the year. In most cases, though, they have resources available depending on more immediate needs within functional departments, which makes estimating the availability of these resources more difficult for new projects.

Example: A construction company that is primarily projectized carries out the construction of commercial buildings. Resources within the organization, such as a piece of heavy equipment and the associated operator, are always allocated to projects requiring that specific piece of equipment. Projects are then scheduled such that certain pieces of heavy equipment are needed on only one project at a time. When finished with one project activity, those pieces of equipment move to the next project on the schedule to perform that same activity.

Alternatives Analysis

As the project manager evaluates each work activity for the resources required and the availability of resources to fulfill each requirement, a compromise might have to be made regarding a critical resource requirement that has to be accomplished by an alternate means. This may mean not having an internal resource available, but simply requiring an external contracted resource. The original callout might have required a more experienced human resource when, upon further evaluation, a lower-level resource that is available can probably accomplish the work. If a work activity requires a specific resource that is more specialized in nature, and the organization does not have this resource, the project manager might have to evaluate the activity for an alternative method using available resources. Alternatives analysis, in most cases, simply requires the project manager to reevaluate project activities for alternatives and to think more creatively about how activities can be accomplished.

Published Data Estimating

Depending on the size of the organization, the publication or listing of all available assets and resources that can be available for activities throughout the organization can be of great use in selecting resources to fulfill work activities. This task is much easier in smaller organizations because available resources are commonly known within the management structure, and the selection and scheduling of these resources involve a simple negotiation with the functional department it is associated with.

Information regarding available resources in larger organizations can be much more difficult to identify because multiple locations can represent resources throughout the country and/or the world. The complexity of logistics and scheduling of these resources to be used at one particular location can introduce constraints on the work activity. It is also more difficult for larger organizations to identify resources that can be used elsewhere in the organization and effectively communicate availability. Publishing information on available resources within an organization is most effective when the organization has locations that are closer in proximity and they typically have access resources that can be easily moved and allocated to projects elsewhere in the organization.

Resource Leveling

As the project manager evaluates a work activity for the resources required, he may determine that given the specific type of work that will be carried out within a defined time frame, the requirement reveals more than one resource is needed. This can present a constraint in the utilization of resources, if resources are to be required elsewhere within the organization or there isn't enough time required to complete an activity given the resources available. This can create two categories of resource constraints:

Time-constrained projects—These projects may have specific work activities that fall within a predecessor/successor relationship that forces the completion of that activity within a specific time frame. The majority of that work activity is to be started and completed on specific dates, and usually, this activity is on the critical path, thus requiring special attention from the project manager for precise execution and completion of the activity. In this scenario, time is at a premium, and resources are obtained for the sole purpose of completing the activity within the specified time frame. This usually results in staying on schedule, but the requirement for extra resources to complete project activities within the specified time frame leads to cost overruns. As an example of a time-constrained work activity and how leveling can reduce individual work load, Figure 6.2 shows how activities I and J have time constraints and have to be completed on those particular days. Extra resources may be added to activities E, F, and G to control the durations of activities prior to activities I and J.

Activities	\multicolumn{13}{c	}{Time-Constrained Activity}
	1	2	3	4	5	6	7	8	9	10	11	12	13
A. Level Ground	■												
B. Foundation Markers		■											
C. Dig Ditches			■										
D. Install Forms				■									
E. Install Sub-Plumbing					■								
F. Install Sub-Electrical						■							
G. Install Rebar							■						
H. Inspection									■				
I. Pour Footings											■		
J. Pour Slab Foundation												■	

Figure 6.2 Time-constrained project

Resource-constrained projects—These projects may have work activities that require specific resources, and constraints occur in the availability of these resources. If key internal resources are identified, they may come with special conditions that do not allow the work activity to be completed within the specified time frame. Therefore, because the focus of this work activity is a specialized resource, the time frame and schedule of the work activity are adjusted to accommodate the resource. Constraints on a resource may include the number of specialized resources available, the limited skill set of human resources available within the organization, limitations in technology or capabilities of a resource, or budgetary restrictions regarding the purchase of external resources. An example of a resource-constrained project is shown in Figure 6.3.

Activities	1	2	3	4	5	6	7	8	9	10	11	12	13
A. Level Ground	█												
B. Foundation Markers			█										
C. Dig Ditches				█									
D. Install Forms					█								
E. Install Sub-Plumbing					█								
F. Install Sub-Electrical								█					
G. Install Rebar								█					
H. Inspection										█			
I. Pour Footings											█		
J. Pour Slab Foundation													█

Figure 6.3 Resource-constrained project

Resource Loading

Resource loading refers to the overall number of resources required for a work activity within a specified time frame. The concept of resource loading allows a project manager to introduce trade-offs or compromises between the numbers of resources used to perform work activities against the time frame allocated to complete the work activities in the overall cost budgeted for a work activity. In many cases, project managers see resource loading as a form of triple constraint. The reason is that the number of resources, cost, and time to completion form interrelated constraints whereby the project manager has to make decisions regarding the best course of action depending on a time-constrained or resource-constrained activity. The general point with resource loading is to try to keep human resources on as much of a normal daily schedule as possible and not require excessive amounts of overtime. Additionally, nonhuman resources should be scheduled as available and not require additional budget for extra resources.

Example: In the creation of a large software product, a component of software code that has to be created falls within a critical part of the overall development of the product. The project manager has only a limited number of software engineers and cannot borrow engineers from other projects. Therefore, he might decide to acquire externally contracted engineers over a specific time frame to complete the work activity on schedule and with the limited internal resources available. For another part of the project, a much simpler component of software has to be developed. A few entry-level engineers who are available from another department can be utilized during this segment of software development to offset the cost of the scheduled engineers working extended hours. An example of loading the resources and having a resource availability problem is illustrated in Figure 6.4.

Resources	1	2	3	4	5	6	7	8	9	10	11	12	13	14	15	16	17
A. Develop Requirements	RS SE	RS SE	RS SE														
B. Design Sub-Module A				SE	SE	SE	SE										
C. Design Sub-Module B				SE	SE	SE	SE	SE	SE	SE	SE						
D. Design Sub-Module C				SE	SE	SE	SE	SE	SE	SE	SE						
E. Test Modules A, B, C												SE TT	SE TT				
F. Design User Interface															SE	SE	
G. Sub-Module Integration														SE	SE TT		
H. Final Test																SE TT	SE TT
Available Resources: RS = 1 SE = 2 TT = 1	RS 8 SE 8	RS 8 SE 8	RS 8 SE 8	SE 24	SE 24	SE 24	SE 24	SE 16	SE 16	SE 16	SE 16	SE 8 TT 8	SE 8 TT 8	SE 16	SE 16 TT8	SE 8 TT 8	SE 8 TT 8

Figure 6.4 Resource loading

The project manager uses resource leveling to the best of his ability in managing the overall scheduling of resources on work activities. This generally is required only on work activities that are either time constrained or resource constrained and require extra attention in managing resources to complete the work activities. Resource leveling can be in the form of shifting resources across activities on the project, adjusting the scheduling of start and stop times for certain activities, or adjusting the number of resources on any particular work activity. Moving activity start times to accommodate limited resources is a form of leveling. This does add to the overall duration of the project and engineers can perform task B in half the time, as shown in Figure 6.5.

Resource Leveling																			
Resources	1	2	3	4	5	6	7	8	9	10	11	12	13	14	15	16	17	18	19
A. Develop Requirements	RS SE	RS SE	RS SE																
B. Design Sub-Module A				SE	SE														
C. Design Sub-Module B						SE	SE	SE	SE	SE	SE	SE	SE						
D. Design Sub-Module C						SE	SE	SE	SE	SE	SE	SE	SE						
E. Test Modules A, B, C														SE TT	SE TT				
F. Design User Interface																SE	SE		
G. Sub-Module Integration																SE	SE TT		
H. Final Test																		SE TT	SE TT
Available Resources: RS = 1 SE = 2 TT = 1	RS 8 SE 8	RS 8 SE 8	RS 8 SE 8	SE 16	SE 16	SE 16	SE 16	SE 16	SE 16	SE 16	SE 16	SE 8 TT 8	SE 8 TT 8	SE 16	SE 16 TT 8	SE 8 TT 8	SE 8 TT 8	SE 8 TT 8	SE 8 TT 8

Figure 6.5 Resource leveling

Resource Requirements Plan

After the project manager has evaluated all resources required to effectively and efficiently complete all work activities, the conclusions from these evaluations need to be captured in a master schedule called the *resource requirements plan*. If the project manager has documented the initial structure of the project and breakdown of each of the work activities by using a work breakdown structure (WBS), information on the resources required should have been captured during the information-gathering process, and specific information on the availability of each resource can now be included. This information can also include specific requirements of the resource itself, notes regarding critical scheduling of particular resources based on availability, and indications of the need for external contracting of resources not available within the organization. The resource requirements plan is then a component of the overall WBS and project management plan. If the project manager does not use a formal WBS structure, a simple matrix listing all resources required for each work activity, specific requirements of each resource including availability, any external contracting of resources required, and scheduling requirements of each resource can be considered the resource requirements plan.

The project manager needs to define the requirements of each resource and the commitment of scheduling resources through as much of the project life cycle as possible at the beginning of a project. If commitments to scheduling resources are not made early in the project, and resources are not available when required, this generally results in either extending the schedule to accommodate resources or the extra expenditures to bring in resources when required on the schedule, resulting in budget overruns. With regard to planning and estimating resources on a project, the project manager can decide one of two things:

1. Ignore planning resource requirements in advance and obtaining commitments and scheduling of all resources.

   Result—Schedule constraints and conflicts that typically result in extending the schedule or having budget overruns due to lack of resources by having to pay a premium for external resources.

2. Spend the time to ensure resource requirements have been properly identified and estimated, and commitments have been made for scheduling of all resources required for work activities.

   Result—Mitigate or eliminate the risk of the unavailability of resources, which can help eliminate schedule slipping and cost overruns due to lack of resources.

In designing and developing the project plan, the project manager must pay close attention to activities early in the project such as defining and estimating resource requirements. Because most projects rely on resources to carry out project work activities, it is incumbent on the project manager to make the evaluation of all resources required on the project a top priority. In most cases, the most critical element to estimating resources is the proper selection of a resource, availability, and commitments for the scheduling of resources as required during the project life cycle. The project manager will find more success in the control of the schedule and cost of a project if adequate time is invested, information is gathered and evaluated, and resources are selected and scheduled as early as possible in the development of a project.

Review Questions

1. In your opinion, do certain project resources hold a higher importance than other resources? If so, why?
2. What are the advantages and disadvantages of using external contracted resources?
3. Discuss some of the primary or more common constraints with project resources.
4. Explain the challenges in project staffing due to human resource skill set availability. How would you overcome these challenges?
5. Discuss some of the differences in estimating methods and why you would choose one over the others.

Applications Exercise

Klanton Bower Data Center: Case Study (Chapter 4)

Apply the concepts described in this chapter to the case study:

1. Develop a list of resources needed on the project.
2. Can you identify any resource constraints for this project?
3. Are there any special capability requirements for resources?
4. Which estimating method would be best for this project and why?

7

Activity Duration Estimating

Introduction

At this point, the project manager and associated project staff have finished gathering information on individual work activities to define what has to be accomplished and the required resources to complete each activity. It is now time to estimate how long it will take to complete the work of each activity. The process of analyzing all the variables that could have influence on how long it will take to complete work activities is called *activity duration estimating*.

Activity duration estimating requires information that defines what has to be accomplished in the work activity and all the resource types that are required. The project manager also needs information on any periphery requirements such as contractual agreements and/or project or organizational constraints that can have an influence on determining the overall time it takes to complete the work activity. Accurately estimating the time to complete a work activity is one of the hardest things the project manager has to accomplish in developing the overall project plan. This is generally based on the number of variables that any activity will likely have that can influence the duration of time required to complete an activity. Inasmuch as the project manager tries to estimate a time duration for an activity, it is common for these durations to be adjusted throughout the project life cycle based on both observable factors and unforeseen factors that can influence project activities. Nonetheless, the project manager needs to use fundamental project management tools to develop the best estimates possible for time durations on work activities.

Some activities may have very definitive information that the project manager can use to define the activity duration quite easily and accurately. For other activities, however, ascertaining the time duration can be more difficult due to any number of variables that can affect this type of analysis. The project manager must always try to make estimates based on real data and avoid the temptation to simply guess. Doing so can cause serious ramifications with scheduling of other successor activities and may create a ripple effect that can ultimately impact the overall schedule and completion of the project. Inaccurate duration estimating can not only affect the estimated completion time, but also can have other effects throughout the project:

- Negative effects in morale for the workers completing work activities
- Negative effects on management's expectations of the project and project manager
- Negative effects on the customer's expectations of the organization's ability to complete a project
- Challenges for the project manager in controlling the project schedule
- The potential to increase costs to the project if contractual obligations include late penalties for missed schedule dates
- Negative effects on the project team's confidence and their overall buy-in to the project objective

The project manager and project staff should spend quality time in trying to make the most accurate estimates possible regarding time durations. With smaller projects that have fewer work activities, in many cases it is easier to develop accurate time durations because the overall scope of work is not complex and the number and types of resources are easily quantifiable. With larger projects that have time spans of several years, estimating accurate time durations can become difficult for work activities further out in the project life cycle where accurate information is at a minimum. At this point, the project manager has to rely on information available at the time of estimating and has to note that more accurate estimating will be required as more information becomes available. This type of estimating is acceptable and common with extremely large projects that have long time spans.

This chapter covers several tools and techniques the project manager can use to estimate time durations for work activities that can include both internal and external influences, constraints, and considerations for integrating predecessor and successor relationships. As with many other areas within the project, developing processes to perform activities such as activity duration estimating is critical to the project manager effectively and efficiently creating a project management plan.

Duration Estimating Methods

Depending on the size and complexity of a project, information gathered to define attributes and characteristics of work activities may reveal there is or is not enough information to accurately estimate the duration of an activity. The project manager must have tools to utilize the information available for each work activity to estimate activity durations. Duration estimating tools have to be capable of assisting in estimating activity durations that have marginal as well as ample information available. In the following sections, we explore several of the more commonly used duration estimating methods and examples illustrating how they are used.

Analogous Estimating

When little or no information is available to characterize the duration of an activity, the next best thing is to compare the characteristics of that activity to an activity of a similar project completed in the past. Historical data from project activities can help the project manager visualize how an activity might play out based on the *analogy* that if a certain activity, given certain parameters, took *X* number of hours, then based on similar parameters, the new activity should also take *X* number of hours. The project manager uses the comparison of activity characteristics and parameters matched to a past activity that would suggest a particular duration to estimate the duration for the new activity. This type of estimating is considered to be a *gross estimation* because it is quantifiable only by a comparison methodology.

Example: A project manager is estimating durations for work activities on a custom software development project. She has discovered that it has been difficult to estimate the overall activity duration for developing the user interface component of the software package. She investigates a software package that has a similar user interface that was developed the previous year. By doing so, she discovers the work activity has similar characteristics to the current project activity. Upon further investigation and comparison, she discovers the activities are similar enough that she can use the duration recorded on the past activity as an estimate for her new project activity.

Parametric Estimating

Parametric estimating is used in conjunction with analogous estimating where historical data has to be used to derive the basis of activity duration but parameters are different. Consequently, calculations must be made using the historical data to formulate the duration for the new work activity. This approach is typically used when the comparison of work activity characteristics of a past project to the present project reveals enough similarities that duration information can be used. Parametric estimating introduces a multiplier component that adjusts for variations in parameters such as size, quantities, or manpower that will scale past data so that it is applicable to the present work activity's parameters. This type of estimating is also considered to be a gross estimation because it is quantifiable only by historical data and mathematical calculations to estimate durations.

Example: John is the project manager overseeing a custom home project and is developing the project management plan. He is estimating durations for several work activities and is struggling with the overall duration for the completion of the foundation. He has not been able to get a definitive answer from the contractor but needs to complete these duration estimates to report an estimated time at completion for the customer. John has managed custom home projects in the past, and based on the layout of this particular home, he identifies a past project that has similar characteristics in shape. The previous home was quite a bit smaller, but the characteristics of the foundation are exactly the same as the proposed home he is working on now. He then takes the square footage of the foundation and, in comparison with the size of the new foundation, calculates the increase in square footage. Next, he uses that multiplier to scale all the other factors such as manpower and equipment to derive an estimated duration for the new foundation. This approach not only gives John a rough estimate of the overall duration to complete the activity, but also provides a better idea of manpower and equipment needed to complete the activity within that time frame.

> Past project foundation size and duration: 1800 square foot, 96 hours
>
> Parametric multiplier calculation: 96/1800 = 0.0533 hours per square foot
>
> New project foundation size: 3200 square feet, hours (not known)
>
> Apply parametric multiplier: 3200 × 0.0533 = 170.56 hours
>
> Parametric estimation for duration of new foundation: 170.56 hours

Three-Point Estimating

In some cases, information gathered for a work activity might reveal several pieces of data that may suggest a range of duration values. In this situation, the project manager may have a dilemma as to which is the right piece of information to use for the absolute value of the duration. He could solicit advice from subject matter experts as to which value should be used or refer to prior projects, but if this type of information is not available, the project manager must use a method to determine an absolute value for work activity duration.

Some project managers may choose to use a more conservative (*pessimistic*) value, whereas other project managers might develop a more aggressive schedule (*optimistic*) of work activities. If the difficulty lies in choosing which one should be used, a method called the *three-point estimating* process allows the use of both optimistic and pessimistic values to calculate an expected duration. Two forms of this method are the *triangular distribution* and the *beta distribution*. The beta distribution form of the three-point estimating method was originally developed as part of the program evaluation and review technique (PERT):

Optimistic (T_o)—This estimate is based on data that would suggest an absolute best-case scenario condition for a work activity. This requires all factors and resources within the work activity operating at peak efficiency and providing the shortest duration possible.

Most likely (T_m)—This estimate is based on data that would suggest a nominal value of duration with little or no risk or constraint influence. If all goes well, this is what is most likely to happen within a project activity given available resources, materials, and a reasonably designed time frame.

Pessimistic (T_p)—This estimate is based on data that would suggest an absolute worst-case scenario condition for a work activity. This can include several factors and/or resources within a work activity having problems, risks, or uncertainties that will influence work activity and a general overall lack in efficiency, resulting in longer-than-normal durations.

Expected duration (T_e)—The expected duration is the mean of the distribution calculations (shown in Eqs. 7.1 and 7.2) taking into consideration all three classifications of estimates: *optimistic, most likely,* and *pessimistic*. The project manager can use this value as a duration estimate outcome for a particular work activity.

Triangular Distribution: $$T_e = \frac{(T_o + T_m + T_p)}{3}$$

Beta Distribution (PERT): $$T_e = \frac{(T_o + 4T_m + T_p)}{6}$$

The beta distribution is similar to the triangular distribution but applies more emphasis to the *most likely* data and deemphasizes the two extremes (*optimistic* and *pessimistic*), as shown in the formula where t_m has a multiplier of 4 and is divided by a factor of 6 to normalize the magnitude of distribution. This allows the project manager to take into consideration the influence of the two extremes, but they will carry less weight in the overall calculation of the duration estimate.

Using the data in Figure 7.1, you can use the beta distribution (PERT) formula to calculate the expected times for each activity on the critical path and derive the overall estimation of duration for the project.

Activity	Optimistic (t_o)	Most Likely (t_m)	Pessimistic (t_p)	Expected (t_e)
A	3	4	6	
B	3	5	6	
C	2	3	5	
D	1	2	3	
E		3		
F		4		
G		2		
H	4	5	7	
I		2		
J	2	4	6	

Figure 7.1 Optimistic, most likely, and pessimistic estimates

You can apply the values of activity A in the beta distribution formula to determine the expected duration, as shown in Figure 7.2. Continue for the remainder of the critical path activities and see how the estimates are affected by the use of distribution.

$$4.17 = \frac{(3 + 4(4) + 6)}{6}$$

Activity	Optimistic (t_o)	Most Likely (t_m)	Pessimistic (t_p)	Expected (t_e)
A	3	4	6	4.17
B	3	5	6	4.83
C	2	3	5	3.17
D	1	2	3	2.00
E		3		
F		4		
G		2		
H	4	5	7	5.17
I		2		
J	2	4	6	4.00

Figure 7.2 Beta distribution calculations of activities A–J

Three-point estimating offers a unique value: It allows the project manager to utilize a range of data that may take into consideration those risks and uncertainties that may impact the duration of a particular work activity. Both optimistic and pessimistic values can reflect the magnitude of variables a project manager might be faced with or constraints that have to be considered for estimating the duration of a work activity.

Contingency Estimating (Reserve Analysis)

Most project managers take the information of each work activity at face value; they then proceed to calculate duration based on what's most likely to occur given the information available for that activity. Because most projects do not go as scheduled, and problems, constraints, and either identified risks or unknowns influence the outcome of work activities, this can present a challenge to accurately forecasting durations for work activities.

Because the project manager spends a great deal of time analyzing the potential risks throughout the project and either qualifies or quantifies the impact improbability these risks might have on the project, he also designs mitigation or elimination strategies and develops contingency plans if risk events are to occur. Contingencies for risk events can be in several forms, such as extra funding, additional resources, or alternate plans, depending on what's required to address a particular risk. In some cases, extra time can be designed into work activities (*reserve analysis*) if a risk with high probability or high impact to the project is identified, and extra time should be calculated into the activity to compensate; this is called *contingency estimating*.

The project manager has two ways to design in extra time on a project for particular work activities that may require special attention:

Activity contingency estimating—If a particular activity is identified with a high-probability or high-impact risk, the project manager may choose to assign duration contingency for that specific activity. This is an extension of time allotted for a specific activity that is calculated into the baseline of the overall schedule of the project, but is used only if the risk event for that activity is to occur.

Project contingency estimating—The project manager can also choose to compile all activity contingency estimates that have been calculated and simply add this as one block of time to the overall project schedule. It is required that this block of time also be calculated into the project baseline based on high-probability and high-impact risk events that have been identified throughout the project life cycle. This is a more difficult approach for the project manager to control; he must keep track of how much contingency time has been used for risk events versus simple inefficiencies of project activities that use up time in the project schedule.

Subject Matter Expert Analysis

If project managers find it difficult to gather accurate information for the duration of a particular work activity, it is advisable they seek the counsel of subject matter experts who can analyze the activities and help determine a duration estimate. In some cases, the project manager may formulate a *group decision-making technique* that brings together individuals with various backgrounds, experience, and knowledge to help him make a decision for a work activity duration value. Due to the complexity of work activities, some projects may require a panel of experts to evaluate the characteristics and parameters of work activities to formulate a consensus as to the estimate of work activity duration. In some cases, the project manager may discover new information about a work activity based on discussions with individuals within the group; this information can shed light as to the reevaluation of the duration of a particular activity. If this group decision-making technique includes individuals identified to perform the work of the activity, this exercise may improve their perception of the importance of staying on schedule and performing the activity efficiently as designed.

Duration Estimating with Constraints

When a project manager is analyzing information gathered for work activities, most of this information points to parameters and characteristics directly related to the attributes of the activity itself. In some cases, the project manager may discover other influences that might affect the work activity duration and must take them into account. Outside influences typically include a condition that imposes a limitation, regulation, or constraint that will force a time duration to be adjusted or held at a fixed value such that there can be no variation or extension. Constraints can be in several forms but typically have the same characteristic in regulating activity duration.

Triple Constraint

The classic project management constraint that all project managers face on just about every type of project is the relationship between cost, time, and quality of the deliverable for a work activity; this is called the *triple constraint* (see Figure 7.3). When a work activity has specific requirements defining what the deliverable will be and the associated cost of resources and materials to produce that deliverable, the third element is the time that it takes to complete the activities required of the deliverable.

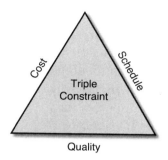

Figure 7.3 Triple constraint

The theory of the triple constraint requires cost, time, and quality of the deliverable to be maintained as designed to accomplish the activity objective on budget and within the time allotted. If one item is altered, such as the activity taking longer than expected, either more resources are required to stay on schedule at a higher cost, or if the cost is constant, the quality of the deliverable is compromised to complete it within the scheduled duration. If any one item is compromised, one or both of the remaining two items in the triple constraint will suffer to maintain the budget, schedule, and the completion of a deliverable that the customer may or may not find acceptable. It is incumbent on the project manager to correctly and accurately estimate these three parameters for each activity to avoid compromise within the triple constraint.

Top Down/Bottom Up

Inasmuch as the project manager can perform the task of managing the triple constraint within a work activity, he also may have to contend with other influences within the project and the organization. Depending on the size and complexity of the project in the structure of the organization, the project manager may find variability in the commitments of resources that were supposed to be available for project activities. If these variabilities are seen at the workforce level, and the availability of human and other resources such as equipment and materials present a challenge, this is considered *bottom-up constraints*. In most cases, just the day-to-day operations introduce variability regarding the availability of resources. For example, human resources may call in sick or be injured on the job, thus delaying work activity; unforeseen weather may not allow work activity; and other basic elements at the work activity level can present challenges to accurately estimating the duration for a work activity.

If influences are directed from a higher level such as functional managers, other management, and executives within the organization, that determines either a different course of action for a work activity or an alternate allocation of resources that can affect estimating a duration for a work activity; this is called a *top-down constraint*. Based on the information of a work activity, project managers may have a clear idea as to a very accurate duration, but upper management may tell them that a different course of action is required that will change the value of the initial estimate for a work activity. There might even be commitments for critical resources for specific work activity that management has to change based on other obligations within the organization, and the project manager has to design a different course of action that may result in a different value of estimated duration.

Customer Requirements

Another common area of influence for work activities can be changes imposed externally by the customer that require a reevaluation for an estimated duration. The project manager, who is typically the point of contact for a customer on the project, has to be mindful of customer changes that will result in altering project schedule and/or budget or possibly parameters or characteristics of the deliverable. It is best to have a change order process in place that documents requested changes by the customer that have to go through analysis and approval processes to properly initiate those changes on the project. This process can help the project manager because changes to the scope of the deliverable that may produce changes in the budget and schedule can actually be adjusted on the baseline and be accounted for officially within the project. This is the best-case scenario because the project manager can actually make changes to the work activity directly to reflect what the new requirements will be.

If this change process is not in place, customer requests for changes can result in incremental alterations to the scope of the deliverable, which can, in turn, incrementally increase cost and extend durations. In this event, staying on the original budget and schedule becomes difficult for the project manager to control; this is called *scope creep*. When the project manager initially makes estimates for durations on work activities, it is especially important that any change requirements imposed by the customer be carefully documented. If they are approved, estimated durations need to be altered to capture new information that may affect the overall outcome of the project. Careful analysis of change requirements may also reveal new constraints that may be present for a work activity that did not exist prior to the change; they should be considered in the approval process for a customer change request.

Successor/Predecessor

When the project manager initially creates the network of project activities, and the successor and predecessor relationships become visible, this information also reveals constraints that these relationships may impose on particular work activities. In the information-gathering process for each work activity, it is important to note whether there are critical start or stop times for particular activities so that these constraints can be identified for a corresponding successor or predecessor activity. This task is vitally important in the initial design of the network so that the project manager can identify potential alternate paths of work activities that may eliminate or mitigate the influence of certain constraints. During the design phase of project activity sequencing, the project manager has the most influence on creating or eliminating successor and predecessor constraints due to the visibility of relationships between work activities, as shown in Figure 7.4.

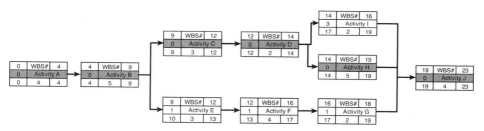

Figure 7.4 Successor and predecessor constraints in a network diagram

As you can see, activities I and H cannot start until activity D is complete. However, it was determined that activity I doesn't need to have D as a predecessor and can be located in a different location in the network. It was also determined that activity D is not dependent on C and can actually be performed in parallel. This information can give the project manager some options in changing some activity relationships and placement in the network that can reduce constraints and improve the overall project schedule, as shown in Figure 7.5.

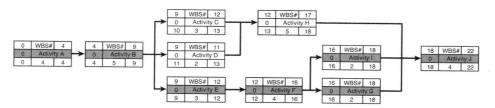

Figure 7.5 Relieve constraints and dependencies for a new network

It is important the project manager identify these constraints early to properly plan for critical timelines as well as the possibility of contingency estimating for certain critical work activities. It is also important the project manager take this analysis seriously for accurate duration estimating based on constraints that activity relationships can present. The process of analyzing relationships of successor and predecessor constraints and the corresponding alterations that can be made within the network is called *duration integration analysis*.

The best time for project managers to make any types of changes is during the design and development stage of the project plan. In most cases, during the development stage, work activities have not been started, and simple alterations to work activity parameters and the structuring of work activities can be made easily if proper analysis has been performed to ensure the most efficient sequencing of activities has been accomplished. It is during this time the project manager can solicit advice from subject matter experts as to the proper sequencing of activities based on critical constraints one activity may impose on a neighboring activity. Networking models can be created and run to show the forward and backward pass analysis that produces the overall project duration and the effects of critical placement of certain work activities that have successor or predecessor constraints. Networking models can also reveal any slack time that activities may have that can be factored into the estimate for a project activity. The project manager should always take advantage of this opportunity at the beginning of a project to utilize tools to understand each work activity and make any alterations, if any, to maximize the efficiency of the overall project.

Scenario Analysis

After the project manager analyzes various critical work activities that have duration constraints using a network diagram and various alterations of estimating duration, he can then analyze different scenarios for project plan development. Some components of the analysis may include slight alterations regarding how activities are actually performed given more or less resources, the availability of more or less budget, and any alteration that can be made in the actual deliverable required by the work activity. It may be possible to combine activities for better utilization of resources and to help mitigate or eliminate the effects successor or predecessor relationship constraints might have. It is always best if the project manager thinks out of the box and tries to be as creative as possible in understanding how to eliminate the influence of constraints on individual work activities and the overall project.

Running simulations for different network diagram structures as well as understanding external influences to the project may reveal better ways to structure project activities to eliminate unwanted stress and constraints. In some cases, this might require soliciting the approval of the customer for slight modifications or alterations in the product deliverable for the elimination of constraints that could be detrimental to the project. Subject matter experts may also shed some light on alternate scenarios for design and/or manufacturing of certain items that may eliminate constraints. The important factor for the project manager is to try to consider alternatives and scenarios and leave no stone unturned as to the possibility to reduce or eliminate either constraints or potential risk for work activities when developing a project plan.

Scheduling Conclusions

In the course of developing the project schedule, the project manager will discover that this process is one of the most difficult and critical in the overall development of the project plan. The project manager also should understand that, depending on the size and complexity of the project and available resources, it is best to have assistance in gathering information, soliciting expert opinion, and analyzing the most efficient sequence of activities. There also are estimating methods that can utilize quantifiable absolute duration values as well as duration estimating with little or no information available.

The process of estimating activity durations can ultimately impact the project at the *activity level* and influence the scheduling of *major milestones* within the project and the allocation of resources across to other *projects and programs* within the organization. Activity estimating incorporates the information specific to an activity but also has to take into consideration influences of available resources within other projects or programs and the effects this can have in estimating activity durations and possibly the sequencing of activities within the project. Inasmuch as we have discussed the details of duration estimating within this chapter, critical elements of duration estimating fall into three fundamental categories:

Information gathering—Information gathering is the backbone of defining activity requirements. The amount, accuracy, and various types of information that can be gathered for a work activity are vitally important. Within this information are the details of the relationships that the specific work activity will have with other activities that may create predecessor or successor relationships and possible constraints. Information on an activity also reveals what type of duration estimating method will be used and the accuracy of that estimate. It is important that the project manager and those assisting him during the information-gathering process be aware of the importance of how much information is gathered and the quality and accuracy of that information.

Duration estimating methods—Based on the information gathered for each work activity, details or the lack of details regarding activity attributes and parameters dictate what types of estimating methods will be more successful in producing duration estimates. The project manager can select an estimating method using absolute parameter data that would suggest well-defined activity duration. If information suggests a variety of both positive and negative influences that could affect the duration, the project manager may choose to use a method that can incorporate a range of data. In some cases, there is simply not enough objective information to draw a conclusion for activity duration, and the project manager has to select an estimating method based on data other than what is present within the information for this particular activity.

Network diagram analysis—After the project manager uses activity information to select a duration estimating method and derives an estimated duration value, he should analyze a network diagram to validate the accuracy of the estimated duration for each work activity when associated with all other activities in the project. This analysis can reveal constraints that have been created based on predecessor and successor relationships as well as the correct sequencing of activities and optional paths that might be available based on estimated activity durations. The segment of the overall estimated duration's process is most critical because it influences the overall design of the project plan and estimated time of completion. This is also the point where the project manager has the most leeway in making changes to the sequencing and connection of work activities within the network to mitigate or eliminate risk based on constraints.

Activity Level

As we have seen, estimating durations has the biggest impact at the activity level, where the utilization of resources and the timing of activities with relation to neighboring activities is of paramount importance. This is the top priority for the project manager in developing the project plan as he sets the initial structure of all relationships that each work activity will have to neighboring activities and the influences those relationships will create. Setting the duration of an activity imposes a time frame in which resources and materials have to be utilized, and the project manager or project staff needs to monitor and control this duration. After network analysis, it may be determined that there is available slack time for activities that may reduce the sense of urgency on some activities. However, this is not the case on all activities, as defined by those activities on the critical path. Critical path activities generally have no slack or have the least amount of slack available, and create constraints to manage resources to complete activities within specified durations. Estimating activity durations accurately is a very important process the project manager performs in the initial development of a project plan.

Project Milestones

In the development of the overall project management plan and the scheduling and sequencing of activities, the project generally encounters stopping points called *milestones*. They require an action before continuing with further project activities after the milestone point. These milestones can be simple stopping points that the project manager has designed in to temporarily halt activities. He must take inventory of what has been completed to confirm project activities are on schedule and on budget and confront any issues that might have to be addressed before continuing. This is a wise thing for the project manager to do, no matter whether there are other milestone requirements, because this allows him to have review points throughout the project.

Milestones can be internal or external to the organization, and they can be designed by the project manager or imposed due to managerial requirements of the organization or external requirements based on the type of project. Some organizations that have a project management office (PMO) in place may require milestones be designed in so that other management staff can have project reviews throughout the project life cycle. Milestones can also be created externally from customer requirements, such as in product development. The reason is that the customer wants to review what has been created at certain points in the project to ensure it is satisfactory to meet the overall project objective. Milestones may also be external based on state, local, or federal regulatory stopping points for inspection requirements. This is common in the construction industry, where inspections are required at certain points in the construction of a building.

Because these milestones may be designed by the project manager or imposed by internal or external requirements, they can create duration constraints and duration estimates that play a role in influencing milestone schedules. If the project manager designs a network of activity relationships and determines the most efficient sequencing of activities, placing milestone points after the fact can create conflicts and constraints because milestones have an activity associated with them but may or may not have a time duration, which in many cases may be undetermined. For example, having a city inspection performed on a foundation may be designed as a one-day milestone event initially. In reality, however, the city inspector may not be able to commit to a particular day for inspection, which can cause problems in scheduling the start of the next activity.

Likewise, the milestone itself can create problems if the project manager inaccurately estimates the action of the milestone and any estimated duration that it may require. For example, in the development of a prototype, a milestone is designed (from internal organizational requirements) partially through the development phase and has no published time requirement, but the customer is not aware that this milestone is in place. The customer expects delivery of the prototype on a particular date, and the internal requirement of this milestone requires review of engineering developments that may cause delays. The requirement of this milestone and unplanned actions as a result of reviews now have this milestone affecting the overall project schedule although it was originally intended to be simply a sign-off function.

Having milestones within the project life cycle is typically a good thing, but it is important they be designed into the project schedule and an estimated duration be associated with each milestone just to give the schedule slack with each milestone if time has to be used. Some milestones may simply be a sign-off function, and no time is required. In most cases, though, milestones, even with a sign-off, may require several hours or even a full day.

Project and Program

Estimating activity durations can have an influence and can be influenced when the project shares resources with other projects and programs within the organization. This fact typically is noted in the initial information-gathering process so that the project manager has this information available when performing estimated durations for each activity. In most cases, if resources have critical schedules based on the utilization across several projects, it is incumbent on the project manager to coordinate with other project managers, a program director, and other functional managers as to the resource availability. This is factored into the estimated duration of a given activity based on the availability of resources and may also influence when activities will be scheduled within the network of other activities.

In most organizations, the project manager should not manage the project in a vacuum and expect all resources to be available as designed by the project plan. Although the project manager may be controlling his project correctly—being on schedule and on budget—other projects may not be as fortunate and may have delays in the completion of work activities that influence the availability of resources for other projects within the organization. The project manager will not know these scenarios are occurring until after the project has begun, but should be aware of certain resources that have critical scheduling requirements so that he can monitor these resources on other projects and be aware ahead of time if delays are imminent. This provides the project manager a warning and the option to make alternate plans if necessary based on updated resource availability.

In some organizations, the program director, in overseeing the bigger picture of several projects, can manage critical resource utilization and may assist in allocation of resources based on constraints in activity scheduling. The program director usually plays a vital role in the influence of resource utilization and allocation across projects within her program because this can affect whether projects stay on schedule and on budget. It is also important for project managers to communicate accurate status of project activities to the program director because this information also influences the availability of critical resources needed on other projects.

As the project manager completes the initial development of his project plan, which generally includes the sequencing of activities within a network diagram, the program director can align the schedules of several projects to ensure resource availability across several projects. At this point, accurate duration estimating is extremely important because the program director will start projects based on the alignment of work activities and the allocation of resources across several projects. It is then important for the project manager to ensure projects stay on schedule because other project managers are counting on resources being available for the success of their projects.

Review Questions

1. Discuss the differences between the various duration estimating methods and identify advantages or disadvantages between them.
2. Discuss the primary benefit in three-point duration estimating.
3. Explain the use of contingency estimating.
4. Explain any constraints in using the top-down versus bottom-up approach in duration estimating.

Applications Exercise

Klanton Bower Data Center: Case Study (Chapter 4)

Apply the concepts described in this chapter to the case study:

1. Select a duration estimating method based on the information available in the case study to derive activity durations.
2. What sources of subject matter experts might be available?
3. Are there any identifiable estimating constraints?
4. Determine activity duration estimates for each activity.

8

Schedule Development

Introduction

Developing a project schedule is the final stage in bringing all the information pertaining to the development of a project deliverable to one final conclusion; this includes the strategy of how and when a project objective will be accomplished. This task typically includes correctly identifying the overall scope of the project objective; customer requirements; breakdown of the project deliverable into work activities; and requirements of each work activity with regards to cost, resources, and scheduling. Depending on the size and complexity of the proposed project and the organizational structure, creating a comprehensive, effective, and efficient schedule of work activities is a large and important task for the project manager to accomplish.

On smaller projects, the project manager may develop a project schedule in a very short period of time due to the small number of activities, requirements, and constraints that have to be considered. In this case, the project manager can move quickly and begin the first project activities. In other cases, objects may be very large and complex and have several hundred work activities; they may include several thousands of human resources as well as other nonhuman resources. These resources may have very complex predecessor and successor relationships that make developing a schedule extremely challenging. In many cases, the project manager might elect to have project staff help gather information and analyze work activities to better understand the sequencing of events before developing a project schedule. Performing complex calculations and analyzing work activities to formulate objective conclusions to resource requirements, costs, and durations of activities may take several months or possibly longer. It is important the project manager understand the critical nature of gathering as much data as possible to quantify the characteristics and parameters of each work activity. This information can be used in the analysis of activity durations that will ultimately be used in the development of the project schedule.

The project manager also must understand the tools and techniques available to analyze work activities and the relationships with other activities that formulate the network of activity connections that define a project schedule. In addition, the project manager must understand the importance of analyzing a schedule after it has been completed to verify it is correct and identify any constraints or risks that will be present throughout the network of activities. This network also reveals the most critical path of activities that have to be managed to control the project budget and schedule. The importance in gathering accurate data on project activities and formulating accurate estimates for work activity durations cannot be emphasized enough because this task plays an important role in the overall development of the project schedule.

This chapter briefly reviews how to gather information for schedule requirements and organizational and resource constraints and requirements. It also covers basic network diagramming methods, schedule development and analytical tools, and the proper documentation and management of a project schedule. The project schedule has to be developed correctly before the organization can begin its first project activity. In many cases, the project manager may have to report the overall estimated budget as well as the estimated completion time before a project can begin. Developing the schedule allows the project manager to arrange all the activities such that she can identify and quantify an estimated value for cost structuring and schedule structuring that will give all stakeholders a more refined idea as to what will be created, how much it will cost, and when it will be completed. Unlike the project charter and scope statement of work that have very rough estimations of cost and schedule, the project schedule at this point should have an extremely refined and accurate estimation of cost and schedule requirements.

Schedule Requirements

With any endeavor that someone sets out to accomplish, certain information needs to be gathered prior to starting an activity. In most cases, some form of constraint or requirement also has to be considered in the completion of that activity. With regard to developing a project schedule, many things have to be considered. This requires understanding all the requirements and constraints for each activity that have been identified in developing a project deliverable. Information related to specific requirements called out in the statement of work or specification for a project deliverable can help the project manager understand certain scheduling requirements, whereas other information may have to be gathered regarding the availability of resources and other scheduling constraints or requirements imposed by the organization or by external influences. While developing a project schedule, the project manager must have the mindset that accurate information is most critical to characterizing the requirements of each work activity and how activities will be sequenced and controlled within the project schedule.

Information Gathering

Accurately developing a project schedule starts with examining the information gathered to define the expectations for the project objective and each work activity required in developing and completing a project deliverable. As stated previously, it is important the project manager pay close attention to how information is being gathered with regard to who is collecting the information, what information is being collected, and from what sources it is being collected. Accuracy of information is important because this is the primary determinant of work activity duration estimating as well as schedule development. One primary component to the accuracy and relevance of information is the reliability of the source of that information. It is important for all those who are gathering information to understand the importance of whom or where that information is sourced from because this can lend to a primary component of the relevance or accuracy of that information. The project manager and project staff also may use the following sources of information:

Project charter—This higher-level document typically has enough rough information to outline the general expectations of a project objective. This document serves primarily to inform upper management of a proposed opportunity, and it serves as a primary document for review and approval of a project and the assignment of a project manager to oversee development of a project plan. Information available within the charter may have been communicated by the customer, determined internally, or agreed upon between the organization and the customer; it might be of use in understanding requirements that may have to be considered in developing the overall project schedule.

Project scope statement—This statement generally has more specific information defining the scope of the project objective and more details about the project deliverable that may shed some light on requirements or details that the project manager may need to consider in developing the master project schedule. The scope statement typically helps the project manager understand the boundaries or requirements of work activities, and this is helpful in estimating activity durations to ensure that work does not extend beyond what was required.

Customer specifications—In some cases, customers might have developed specifications identifying in more detail the characteristics, attributes, and parameters of particular parts of a project deliverable. This type of document can be very useful for the project manager in understanding not only the scope of what has to be completed, but also what resources are required. Based on the availability of resources, the project manager can estimate activity durations.

Work breakdown structure—If the project manager has elected to document specific work activities in the form of a work breakdown structure (WBS), it can be useful in determining activity durations as well as providing an initial visualization of how activity sequencing will be determined. The work breakdown structure is also a valuable tool in providing information for network diagramming of work activities.

Specific work activity information—After breaking down a project deliverable into its smallest elements called work packages, the project manager needs to determine requirements of resources, materials, cost, and time duration for each work package activity. After the project staff have gathered information for each work activity, this information is the primary source the project manager reviews in developing a project schedule because it typically has the most detailed and accurate information available for each work activity.

Organizational requirements—Because the project manager first has to focus primarily on specific activity information in developing the project schedule, other factors outside the project and within the organization may impose challenges or constraints that she must take into account in developing the schedule. The organization may have certain policies and procedures in place that the project manager has to factor into the allocation and scheduling of resources. If the project is one of several within a program, the program director might have to approve certain resources and/or scheduling of work activities based on other project work activities that will be carried out using the organization's resources. In some cases, the availability of finances through the organization may present limiting factors for some work activities and may impose scheduling constraints or challenges based on the availability of finances to fund activities.

External influences—Depending on the size and complexity of a project, external factors such as the availability of contracted resources, regulatory requirements, weather, and other external influences may present challenges for the project manager in developing a project schedule.

Project Scope

When developing a project schedule, the project manager should make a practice to step back and take a fresh look at each of the project activities. She must make sure that the scope of work for each activity still falls within the requirements of the overall deliverable and objective of the project. A common mistake that project managers make is that after they do all the work of breaking down the deliverable into its smallest components, and information gathering has resulted in large amounts of data characterizing each activity, some work activities can expand in scope and require more resources and time to complete. This might be the result of customer requests to include more items within specific work activities that require longer durations, or it may be the result of the project management staff overdeveloping a work activity that now requires more work than is actually needed. The project manager must be careful that all work activities stay within the scope of what they were originally intended, so as to help manage the estimating of activity durations that can influence the overall development of the project schedule.

Resource Requirements

Because most projects require resources to perform work activities and various resources throughout the organization to complete project objectives, the project manager has to be mindful of certain limitations that the organization might have in the availability and utilization of resources for projects. Smaller organizations simply may not have some resources required for project activities, so the project manager has to contract external resources to fulfill those requirements. Organizations that have ample supplies of resources may have certain specific usage requirements, such as scheduling availability, transportation of resources from one facility to another, and any authorized user requirements for equipment.

In many cases, the project manager has to solicit external contracted human resources. Doing so requires evaluation of skill set, salaries, and other options within the contract before those resources are available for use on project activities. As with contracts of external resources for equipment, contracts with human resources should not be taken lightly and/or approved by the project manager. Instead, they should be reviewed by someone in the organization who has contract writing and negotiating skills. This task needs to be accomplished before activities can be scheduled and, in some cases, may impose constraints or challenges regarding availability of resources.

Customer Requirements

Because most projects have customers of some kind that require a project deliverable, the customers may not always understand all the requirements for the deliverable in the first negotiations of the project objective. Customers may require changes throughout the project life cycle, and those changes have to be managed through a change request process that may have an impact on work activity durations. In the initial development of the schedule, it is best if customers identify as many changes as possible so that the project manager can incorporate any changes that will have an effect on durations. Customers may have other requirements also, such as the scheduling of critical inspections or milestones that may be difficult to quantify duration to complete.

In some cases, customers may negotiate involvement within certain project activities that require special scheduling considerations and arrangements; this involvement has to be managed carefully during the project. Customers may also require the delivery of partial work activity for review and testing. The scheduling of this activity may be difficult to quantify for activity duration. If customers have an activity scheduled with a specific time to deliver any project components back to the organization, it is typically the project manager's responsibility to ensure timely delivery to stay on the project schedule. In some cases, the project manager may include a buffer in the schedule for activities like this to help eliminate schedule delays during the project.

Schedule Structuring Techniques

Depending on the size and complexity of a project, the project manager needs to choose a fundamental structuring for project activities that allows her to schedule activities within a sequential form and monitor and control work activities to stay within a specified schedule. On some projects, several activities may be accomplished simultaneously, whereas other projects might require a predecessor relationship in which one activity has to be completed before the next activity can be started. Structuring these project activities can take different forms, depending on how the project manager wants to document the sequence of activities.

In choosing the type of structure to document work activities, the project manager also needs to consider how work activities will be communicated to other project staff and resources engaged in those activities. If structures are confusing and difficult to understand, this may cause more problems in trying to document a structure of activities. Structures must follow a logical flow in the way activities are carried out so that the project manager and other project staff can clearly understand the sequence of activities. If the project manager was involved in the initial discussions regarding a project deliverable and how that deliverable might be broken up into work activities, there may be a significant amount of time lapse between that discussion and when the project manager is structuring a project, so details regarding specific sequencing of activities may have been lost. The project manager must develop a structure that accurately represents sequencing of activities that can be easily understood and communicated.

Schedule Structures

The project manager has to consider several parameters when deciding what type of schedule structure to use for a particular project. The size and complexity of a project might determine how many individual work activities are required, and overall anticipated duration of the project may also factor into what structure to use. Depending on the type of project deliverable and the required work activities to produce that deliverable, two primary structures are used in project management: activity disposition and activity hierarchy.

Activity Disposition

The activity disposition structure is used primarily on simpler projects that have fewer work activities and is best used when work activities can be done simultaneously. The general philosophy of this structure is that work activities are scheduled as resources become available, and activities will be compiled at the end of the project to complete the overall objective.

An example of activity disposition structuring is a procedural development project in which several components of a procedure need to be written and developed simultaneously; at the end, it will be compiled into a master document that defines the process. This structuring is shown in Figure 8.1.

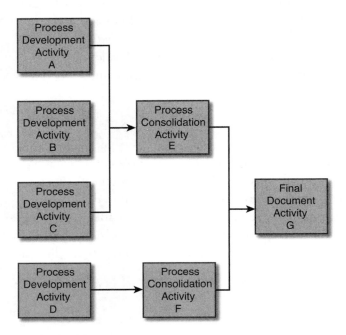

Figure 8.1 Activity disposition structure

Activity Hierarchy

The activity hierarchy structure, more commonly known as the work breakdown structure, is used in large and small projects with varying degrees of complexity that have a combination of independent activities and activities with predecessor and successor relationship requirements. This structure starts with the primary deliverable; it then documents how components are successively broken into smaller segments until the smallest work activities become visible. This structure has an interconnecting attribute that defines the sequencing of activities and also defines predecessor and successor relationships between activities.

You can see an example of this structure in the construction of a house. The completed house is shown at the top and is broken down into large subsections such as foundation, walls, roof, and so on; within subsections are smaller components of actual work activity. It is also evident that certain work has to be completed before other work activities can start, as shown in Figure 8.2.

1.0		Build House
1.1		Initial Work
	1.1.1	Develop Plans
	1.1.2	Get Permits
	1.1.3	Secure Funding
1.2		Foundation
	1.2.1	Level Ground
	1.2.2	Foundation Markers
	1.2.3	Dig Ditches
	1.2.4	Install Forms
	1.2.5	Install Sub-Plumbing
	1.2.6	Install Sub-Electrical
	1.2.7	Install Rebar
	1.2.8	Inspection
	1.2.9	Pour Footings
	1.2.10	Pour Slab Foundation
1.3		Framing
	1.3.1	Frame Walls
	1.3.2	Install Roof Trusses
1.4		Rough Electrical
	1.4.1	Install Main Panel
	1.4.2	Set Boxes
	1.4.3	Pull Wire
1.5		Rough Plumbing
	1.5.1	Install Sewer Drains & Vents
	1.5.2	Install Copper Lines

Figure 8.2 Activity hierarchy structure

Network Diagraming (PDM, CPM)

After a scheduling structure is developed and work activities are defined with information used to estimate duration, an analysis needs to be performed regarding the proper sequencing of activities and attributes of special relationships between activities. This analysis can be performed using a *network diagram*. The network diagram is used primarily for a visual representation illustrating how work activities are connected in sequence to each other. Based on certain requirements, some activities are connected as predecessor or successor requirements, whereas other activities may be shown on different paths as being performed simultaneously with other activities. With estimated durations labeled on each activity, an analysis can be made to show the overall duration of the project, any slack available on each activity, and the identification of a critical path. The network diagram is also useful in analyzing different scenarios in which activities can be moved in sequence or on other paths that might form a more logical and efficient succession of activities to complete the overall objective. This process is called *schedule optimization* using a network diagram.

Network diagramming is commonly performed using the critical path method (CPM) in which activity relationships and connections, along with activity durations, are used to calculate the overall duration of a project. When the forward and backward pass procedures are performed, available slack is identified for each activity in the path. Identifying each path through the network and adding the durations reveals the longest path through the network, called the *critical path*. As explained previously, this process starts with an activity dependency matrix (see Chapter 5, "Activity Sequencing") defining durations and predecessors that serve as the basis for creating a network diagram that can be used in scheduling activities and resources. Based on information from Figure 8.3, the basic network diagram shown in Figure 8.4 identifies work activities in a serial (in-line) dependency connection and the critical path.

WBS #	Activity	Duration
1.2.1	A	2
1.2.2	B	1
1.2.3	C	1
1.2.4	D	1
1.2.5	E	2
1.2.6	F	3
1.2.7	G	2
1.2.8	H	1
1.2.9	I	1
1.2.10	J	1

Figure 8.3 Activity dependency matrix (serial)

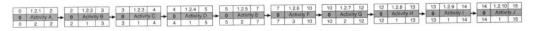

Figure 8.4 Critical path method (CPM) network

Network diagrams with activity relationships that include dependencies requiring predecessor or successor connections use diagramming similar to the critical path; this is called the *precedence diagramming method* (PDM). This approach takes the critical path method of networking and incorporates the requirements of predecessor connections that may alter some of the pathways within the network based on activity dependencies and requirements. Figures 8.5 and 8.6 offer an activity dependency matrix and corresponding network diagram showing the same work activities as Figures 8.3 and 8.4; in this case, though, after some of the dependency relationships are reevaluated, it is determined that the network can be altered to include both serial and parallel predecessor requirements. Precedence diagramming allows the project manager to evaluate all activities in the network for resource and schedule duration optimization.

WBS #	Activity	Predecessors	Duration
1.2.1	A	N/A	2
1.2.2	B	A	1
1.2.3	C	B	1
1.2.4	D	C	1
1.2.5	E	C	2
1.2.6	F	C	3
1.2.7	G	D, E, F	2
1.2.8	H	G	1
1.2.9	I	H	1
1.2.10	J	I	1

Figure 8.5 Activity dependency matrix (serial and parallel)

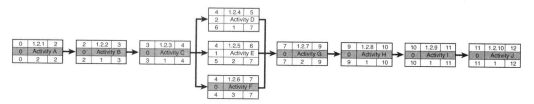

Figure 8.6 Precedence diagramming method (PDM) network

Theory of Constraints (TOC)

As the project manager further analyzes and develops the project schedule, it becomes apparent that certain activities are more critical to manage than others, based on those activities connected along the critical path. Sometimes project managers and/or functional managers may be forced to push human resources harder to ensure tasks are completed on time or to delay the start of projects to adjust to human resource availability. In this case, project managers may also elect to alter the scope of work activities, which affects the product quality, or put more resources on an activity at a higher cost to the project. To avoid this approach, project managers may elect to use *critical chain project management* (CCPM), which results in three activities that may have either positive or negative effects on the overall project:

- Establishing multitasking of resources
- Evaluating task estimates
- Overseeing individual human resources to complete work activities

Ultimately, all projects have various constraints that require project managers to evaluate activity relationships within the network, or evaluate a constraint itself and determine a possible solution to mitigating or eliminating it. The first step in addressing activity constraints is identifying them using the critical chain method (CCM). Subsequently, project managers default to making note of which activities on the critical path need special care in managing, but there may be another way to mitigate the risk of schedule overruns for certain critical path activities.

Eliyahu Goldratt developed the *theory of constraints* (TOC) for production environments and first published this concept in his book *The Goal* (North River Press, 1984). Goldratt identified five areas that deal with constraints and the overall identification and possible elimination of the most critical constraint:

1. **Identify the system constraint.** Identify the primary constraint and root cause of why the constraint exists.

2. **Exploit the system constraint.** After identifying the primary constraint, evaluate other activities within the network of the project for the effect this constraint will have throughout the network. This evaluation may also have to be completed at the organizational level because this constraint may have a ripple effect throughout the organization.

3. **Subordinate everything else to the system constraint.** If the constraint is unavoidable, all activities within the network of the project have to submit to the conditions of this primary constraint.

4. **Elevate the system constraint.** Perform an evaluation within the organization that may identify possible solutions to eliminate the constraint. This results in the constraint no longer holding a primary influence in scheduling of work activities within the network.

5. Reevaluate for a new system constraint. If the elimination of the primary constraint is possible, it is likely another constraint within the project activities may surface as a new point of concern, and the preceding process steps will be applied to the next constraint.

Goldratt used TOC in the production environment to show managers and executives how to focus on the impact that constraints can have within an environment of work activities. These constraints are typically seen as bottlenecks in a production environment, but they have a bottleneck effect on projects, especially activities that are on a critical path. Constraints within a project network can be created based on activity dependencies and predecessor relationships. The constraints can also be found within the activities themselves that may create schedule challenges. The use of TOC in project management is called the critical chain method (CCM).

Critical Chain Method

As the project manager and project staff develop the initial schedule using a network diagramming method, and upon further analysis using forward and backward pass methods of calculating overall project duration and individual activity slack, it may become apparent that certain activities may require buffers or padding to mitigate or eliminate potential scheduling conflicts. Constraints within activities might be created based on the types of resources or availability of resources required for an activity, and the use of schedule buffering may be a way project managers can relieve some of the influence of certain constraints. Goldratt suggested that padding is sometimes used in individual activity duration estimating, so those responsible for the activity have more than 80–100 percent probability of completing the activity on time. In reality, the median time of a 50/50 probability chance of completion indicates a buffer of 30–50 percent. Goldratt offered some theories to explain why this type of padding is included:

- **Parkinson's Law**—Workers use all the time allocated for an activity regardless of how long it actually takes to complete the task.
- **Self-protection**—Workers are reluctant to finish tasks early so as not to communicate their efficiency and set precedents for future tasks.
- **Dropped baton**—An activity is not ready to start and the preceding activity finishes early, resulting in lost time between activities.

- **Excessive multitasking**—Because multitasking is typically encouraged in resource optimization, the excessive use of this tool actually increases work activity duration.
- **Resource bottlenecks**—Constraints in the availability of critical resources can result in delays.
- **Student syndrome**—Procrastination in starting activities can use up valuable slack built into the activity duration.

As you can see, buffers can be designed into work activity durations for several reasons. Because buffers are generally used to manage risk in activity completion, they can be used more efficiently and strategically placed within a network; this is called the critical chain method. The basic concept behind CCM is to utilize the 50/50 rule to lower actual activity durations by 50 percent and redistribute the buffers in the network. As shown in Figure 8.7, each activity is reduced to the 50 percent point, and the buffer is put at the end to manage any uncertainty. Figure 8.8 shows the original work activity durations and critical path. Figure 8.9 gives an example of critical chain method buffering.

Activity	Original Estimates	Adjusted Estimates 50%
A	6	3
B	10	5
C	8	4
D	4	2
Total	28	14
Add Buffer	N/A	14
Total	28	28

Figure 8.7 Activity adjustments 50 percent rule

Figure 8.8 Original critical path

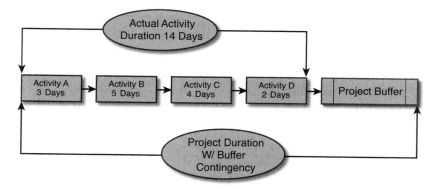

Figure 8.9 Critical path method buffering

Network buffers called *feeder buffers* can be used in the noncritical paths. Buffers used on the critical path are called *project buffers,* as shown in Figure 8.10.

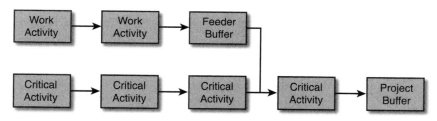

Figure 8.10 Critical chain method buffering, multipath

Schedule Analysis

The initial scheduling of work activities using a network diagramming method is valuable for identifying the correct sequencing of activities and the connection of activities based on dependency relationships. Consequently, the network diagram can also be used to analyze the performance of the project. Performing the forward and backward pass process is the first step in calculating the overall duration of the project, locating available slack on work activities, and identifying the critical path. This information can then be used as the foundation for further analysis of managing constraints, adjustments to resource allocation and the general performance of the project based on the selection of activity placement within the network. The project manager should use various scenario analyses to determine that a better sequencing of activities would result in better project performance by reducing constraints and the overall duration of the project.

As the project manager uses the network diagram for evaluation, it may also become evident that certain activities require adjustments in resources, whereas other activities may require a reevaluation of the estimate in duration. Other tools can then be used to make modifications to project activities with regard to resources, schedule, variabilities, and the optimal placement of activities and allocation of resources within the network. The first area of concern is in what types of resources are needed on each activity and where each activity is in relation to other activities that require the same types of resources; this is called *resource loading*.

Resource Loading

When the project manager refers back to each work activity and the resources required, each type of resource needs to be scheduled for each activity throughout the project network. As resources are scheduled or "loaded" into the project schedule, the project manager needs to analyze how many of each type of resource are required at each work activity. This can also be stated in number of hours of work activity per resource type. If each work activity requires specific resources only for that activity, the project manager has an easier job because this requires only a number of resources based on the availability of resources and hours required on each activity. Difficulty can arise when the same type of resource is required in multiple areas of the project and, in some cases, parallel activities that present a constraint for resource allocation based on limited availability. The simple scheduling chart shown in Figure 8.11 illustrates typical resource loading, and the corresponding network diagram of activity sequencing is shown in Figure 8.12.

		Prototype Telecom Project - Resource Loading				
Activity	Description	Research Scientist (RS)	Electrical Engineer (EE)	Mechanical Engineer (ME)	Assembler (AS)	Test Technician (TT)
A	Develop Requirements	3 Days	3 Days			
B	Design Housing Components			5 Days		
C	Design Sub-Assembly A		4 Days			
D	Design Sub-Assembly B		8 Days			
E	Design Sub-Assembly C		8 Days			
F	Assemble and Test B & C		1 Day		1 Days	1 Days
G	Final Assembly		1 Day		2 Days	
H	Final Test		1 Day			2 Days

Figure 8.11 Schedule of resource loading

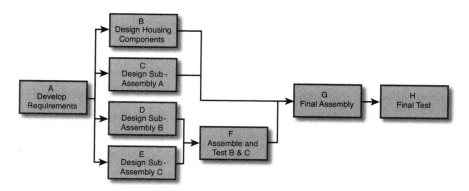

Figure 8.12 Network diagram of activities

It is determined there is an overallocation of electrical engineering resources, creating a constraint for activities to be completed as scheduled with the current resources. As shown in Figure 8.13, too many engineering hours are required in a short period of time, so more engineering resources may be required to complete tasks in the time allowed; or using the same amount of resources, the project manager can add more time to complete tasks.

Resource Loading Requirement																	
Resources	1	2	3	4	5	6	7	8	9	10	11	12	13	14	15	16	17
A. Develop Requirements	RS EE	RS EE	RS EE														
B. Design Housing Components				ME	ME	ME	ME	ME									
C. Design Sub-Assembly A				EE	EE	EE	EE										
D. Design Sub-Assembly B				EE	EE	EE	EE	EE	EE	EE	EE						
E. Design Sub-Assembly C				EE	EE	EE	EE	EE	EE	EE	EE						
F. Assemble & Test B & C												AS EE	TT				
G. Final Assembly														AS EE	AS		
H. Final Test																TT EE	TT
Available Resources: RS = 1 EE = 2 ME = 1 TT = 1 AS = 1	RS 8 EE 8	RS 8 EE 8	RS 8 EE 8	ME 8 EE 24	ME 8 EE 24	ME 8 EE 24	ME 8 EE 24	ME 8 EE 24	ME 8 EE 16	ME 8 EE 16	EE 16	AS 8 EE 8	TT 8	AS 8 EE 8	AS 8	TT 8 EE 8	TT 8

Figure 8.13 Overallocated resources

If it is determined that there is an imbalance of resource availability or resources scheduled on multiple work activities, it is necessary to make adjustments in either the scheduling of resources, the amount of resources required within a specific time frame on multiple activities, or the arrangement of activities. These types of adjustments are called *resource leveling*.

Resource Leveling

One primary advantage in the development of network of activities is a better visual representation of the resources required for each activity as a function of allocation over time. It may be determined that constraints have been created based on the availability of resources and when activities start and finish dates occur. An evaluation may reveal an adjustment, and start and stop dates for certain activities may allow for resources to be utilized more efficiently.

Resource leveling is necessary when similar resources are required for multiple activities on a given project or may be required across several other projects within the organization. Typically, the constraint is in a limited amount of a particular resource, so the project manager has to design the timing of resource utilization across several activities or projects to manage these resources efficiently. To utilize resource leveling, the project manager must go back to resources identified with activities on the critical path because this is where adjustments have the greatest impact to the project schedule. Resource leveling also requires the availability of alternate scheduling of resources because the constraint is typically the result of a resource being scheduled for two or more activities within the same time frame. In some cases, the resequencing of activities may result in altering the critical path, sometimes making it longer, creating a new critical path, or possibly creating a second critical path. With information from Figure 8.13, two days have been added to the schedule. This change allows activity C to be completed before D and E, thus leveling the electrical engineering resources shown in Figure 8.14.

Resource Requirement after Leveling																			
Resources	1	2	3	4	5	6	7	8	9	10	11	12	13	14	15	16	17	18	19
A. Develop Requirements	RS EE	RS EE	RS EE																
B. Design Housing Components				ME	ME	ME	ME	ME											
C. Design Sub-Assembly A				EE	EE														
D. Design Sub-Assembly B						EE	EE	EE	EE	EE	EE	EE	EE						
E. Design Sub-Assembly C						EE	EE	EE	EE	EE	EE	EE	EE						
F. Assemble & Test B & C														AS EE	TT				
G. Final Assembly																AS EE	AS		
H. Final Test																		TT EE	TT
Available Resources: RS=1 EE = 2 ME = 1 TT = 1 AS = 1	RS 8 EE 8	RS 8 EE 8	RS 8 EE 8	ME 8 EE 16	ME 8 EE 16	ME 8 EE 16	ME 8 EE 16	ME 8 EE 16	EE 16	EE 16	EE 16	EE 16	EE 16	AS 8 EE 8	TT 8	AS 8 EE 8	AS 8	TT 8 EE 8	TT 8

Figure 8.14 Resource leveling

Resource smoothing is another form of addressing resource utilization by using available slack within certain work activities. If resource utilization determines that conflicts exist on certain project activities, the project manager can slightly alter resource scheduling based on the requirements of that resource starting work on an activity. If available slack within an activity allows the resource start time to be altered slightly, resource start and stop times can be slightly modified to have a smoothing effect in resource utilization across project activities. The network diagram can be used to identify areas where available slack may allow resource smoothing based on the adjustment of start and stop times. This method can be utilized with little or no effect on the critical path and/or overall duration of the project.

Schedule Reduction Analysis

Upon analysis of the network diagram, it may be determined that the overall estimated project duration is unacceptable and needs to be shortened. Many times, this results in modifications to the scope of the deliverable and/or the procurement of more resources to shorten durations that increase cost to the project. When project managers have to evaluate scheduled activities for reduction in duration, they use two primary tools to reduce the overall schedule:

Schedule crashing—This technique is used to reduce the schedule duration by increasing resource allocation for the least impact to the budget. This technique focuses on particular areas within certain activities on the critical path that can be addressed to reduce schedule duration, such as paying extra for expedited shipping or paying extra for additional resources. Each activity has an associated increase in cost based on the proposed schedule reduction and is evaluated against other activities for the amount of time reduced and associated cost for the reduction. The idea is to select activities with the highest reduction for the lowest cost as an alternative to reducing the overall schedule. The process starts with an activity dependency matrix of project information, including predecessors, durations, activity costs, and estimates on how much each activity can be reduced in duration and cost incurred, as shown in Figure 8.15.

Task	Predecessor	Normal Time	Normal Cost	Crash Time	Crash Cost
A	~	6	$1,200	5	$1,400
B	A	4	$800	3	$1,000
C	B	6	$900	4	$1,200
D	~	8	$1,400	6	$1,750
E	D	6	$900	5	$1,050
F	E	7	$1,400	5	$1,600
G	C, F	12	$2,400	9	$3,000
Total	Critical Path = D, E, F, G	33	$9,000		

Figure 8.15 Activity dependency matrix

Based on the information in Figure 8.15, the project manager can create a network diagram and establish the critical path, as shown in Figure 8.16.

Path #1: (A, B, C, G)   6 + 4 + 6 + 12 = 28
Path #2: (D, E, F, G)   8 + 6 + 7 + 12 = 33

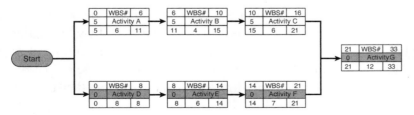

Figure 8.16 Network diagram with critical path

The next step is to determine which activities can be reduced in duration and at what cost. The key in this step is to reduce duration for the lowest cost possible. To find the lowest cost solution, the project manager must calculate the crash cost per day for each activity in the network, as shown in Figure 8.17.

Task	Predecessor	Normal Time	Normal Cost	Crash Time	Crash Cost	Crash Cost per Day	Crash Cost for Critical Path	New Critical Path Time
A	~	6	$1,200	5	$1,400	$200		6
B	A	4	$800	3	$1,000	$200		4
C	B	6	$900	4	$1,200	$150		6
D	~	8	$1,400	6	$1,750	$175	$350	
E	D	6	$600	3	$900	$100	$300	
F	E	7	$1,400	5	$1,800	$200	$400	
G	C, F	12	$2,400	8	$3,200	$200	$800	12
Total	Critical Path = D, E, F, G	33	$8,700	22			$1,850	28

Figure 8.17 Activity dependency matrix with crash cost per day

As you can see in Figure 8.17, the reduction in days can be different, as can the crash cost per day for each activity. If all the reduction is taken on the critical path only, this does not always reflect the lowest cost solution and can create a new critical path. If a combination on the critical path reduces duration equal to another path, at that point, the project manager uses a reduction common to both paths for further reduction. An evaluation of different combinations of reduction and crash cost must be made to determine the "lowest cost" combination.

A sample scenario is to reduce the project from 33 days to 26 days:

Solution #1: D – 2 $175 x 2 = $350   #2   D – 2 $175 x 2 = $350
              E – 3 $100 x 3 = $300   E – 3 $100 x 3 = $300
              F – 2 $200 x 2 = $400   G – 2 $200 x 2 = $400
              C – 2 $150 x 2 = $300   26 $1,050
              26            $1,350

Crashing activities is an effective way to reduce activity or overall project duration, but does require additional cost that needs to be added to the budget. In some cases, activities can be managed through contingency funds that may be available for risk events or uncertainty.

Fast tracking—This technique is used to reduce the project's overall duration by identifying activities on the critical path that were connected in serial but, after evaluation, can be performed in parallel, thus reducing the duration of that path. On closer evaluation of specific work activities, the project manager may determine that the dependency may not exist for one activity; in that case, she may be able to reposition it in the network in parallel with other activities on the critical path, thus reducing the overall duration of the project by the duration of that activity. An example of fast tracking is shown in Figures 8.18 and 8.19.

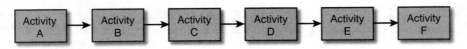

Figure 8.18 Standard serial network

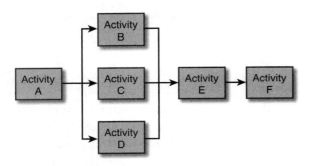

Figure 8.19 Fast tracking a network

Scenario Analysis

After developing the network diagram and reviewing dependency requirements for work activities, project managers may find alternatives that allow for various placements of activities within the network that could produce a more efficient project schedule. One of the benefits of creating a network diagram at the beginning of a project is the availability to run various project planning scenarios that can be evaluated for effectiveness, resource utilization, and overall project efficiency; this process is called *scenario analysis*. In some cases, it requires the project manager to think outside the box for the normal sequencing of activities that might be typical for a particular type of project. Scenario analysis is successful only when completed using different activity sequencing scenarios, and this requires the project manager to be creative in developing various scenarios. Examples of scenario analysis are shown in Figures 8.20 through 8.22.

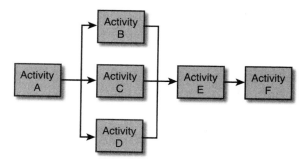

Figure 8.20 Scenario A analysis

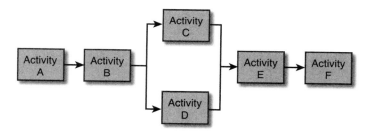

Figure 8.21 Scenario B analysis

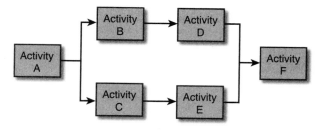

Figure 8.22 Scenario C analysis

Schedule Variance Analysis

Using information from activities on a network diagram, the project manager can analyze the amount of variation in the schedule that certain activities may have on the critical path. This important evaluation gives the project manager an idea as to the possible variations of expected completion time depending on the location of activities in the network and the effect that variations in activity duration can have on the critical path. Schedule variance analysis uses components of the program evaluation and review technique (PERT) as the basis of determining variances for each activity. The calculation for variance of an activity is shown in the following equations.

Formula for three-point estimating (PERT):

$$\text{Expected Time} = \frac{(\text{Optimistic Time} + (4 \times \text{Most Likely Time}) + \text{Pessimistic Time})}{6}$$

Calculate variance for activity duration:

$$\text{Variance} = \frac{(\text{Pessimistic Time} - \text{Optimistic Time})^2}{6}$$

The variance (σ^2) for an activity is

$$\sigma^2 = \frac{(P - O)^2}{6}$$

The project manager may find that information on the variance of each activity is informative but not as useful as anticipated. The best reason to know variances for each activity is to determine the overall standard deviation of the project. The project standard deviation is not the sum total of all activity deviations, but needs to be calculated from the sum of activity variances, as shown in the following formula:

$$\alpha_p = \sqrt{\sum (\text{variances on the critical path})}$$

Schedule Documentation

After the project manager develops a network diagram and analyzes various scenarios to determine the best-case scenario for project activities scheduling, she needs to document the outcome of this scheduling. Depending on the type of schedule structure the project manager uses, she can use various tools to document the information of the project structure and scheduling.

Storage and Software Tools

If an activity disposition structure is being used for a simple or small project, this information can be documented on a software spreadsheet similar to Microsoft Excel that can be stored easily on a network and viewed by others on the project. The software tool can also be used to make comments or notations about various work activities as well as calculations of time and cost structures for all activities listed on the structure.

If the activity hierarchy structure (WBS) is used to document a much larger and more complex project, although it also can be documented in a software spreadsheet format, it is advisable that software such as Microsoft Project be used instead. Microsoft Project is capable of many project management functions, such as automatically displaying the properties of a work breakdown structure in the form of a network diagram. This capability can also make running scenario analysis and crashing exercises simple and quick. Many software platforms are capable of documenting large and complex projects that are part of larger enterprise systems. Therefore, the project manager should solicit information from managers familiar with this platform regarding the availability of a project management module that can be used to document and store project structuring information. It is important the project manager utilize a software platform of some kind to document the structure of work activities so they can be stored for ongoing use by the project manager or other management staff who need access to that information.

Schedule Management

It is important for the project manager to have a properly documented activity schedule that can be accessed on a regular basis because this is how she monitors and controls project activities. The project manager largely uses all the best planning and preparation at the beginning of the project to document what activities have to be performed, what resources will be utilized, and what time frames activities have to be completed within. She can manage these resources and activities to completion only when she has access to a well-documented project schedule. Much of the information available within the project structure and in the network diagram is used on a daily basis to ensure project activities are being performed correctly and efficiently.

Schedule Communication

Projects within an organization have a unique attribute in that they involve many people to either conduct the activities of the project or perform supporting functions to project activities. The project manager and those documenting the project structure can also use the project schedule to communicate project activities to all resources required throughout the project life cycle. If the project structure is documented on a software platform and can be saved within a network location, others within the organization, which may include other facilities, can have access to project information. In some cases, this tool may enable the project manager to announce when activities are starting and when they have completed releasing resources within the organization. The project manager can also use this document to report status to upper management for various project activities. When projects within an organization that utilize resources such as capital equipment, facilities, and financial resources are conducted, the project manager must be responsible for effectively documenting and communicating all project activities. Ineffective communication, in many cases, can be the root cause of schedule delays and the ultimate failure on work activity completion, which is unfortunate if a project has been well thought out and structured.

Review Questions

1. Discuss why it is important the project manager understand the project scope and how that applies to developing the project schedule.
2. What is the significance of customer schedule requirements in developing a project schedule, and what influence can that have in scheduling work activities?
3. Explain the difference between resource loading and resource leveling.
4. Discuss how variations in activity durations may influence paths through a network and could possibly change which one is the critical path.

Applications Exercise

Klanton Bower Data Center: Case Study (Chapter 4)

Apply the concepts described in this chapter to the case study:

1. Review resource requirements and develop a resource loading matrix.
2. Analyze work activity relationships using a network diagram and determine a critical path.
3. Does any resource leveling need to be addressed?
4. Are there other scenarios of predecessor relationships that may improve the network schedule?

Part 3

Project Cost Analysis

Chapter 9   Cost Estimating 185

Chapter 10   Budget Development. 201

Plan Cost Management

Much like developing the schedule for a project, understanding the areas of cost within a project and developing a plan as to how costs will be defined and documented are also large components of the project manager's role at the beginning of a project. These tasks require the development of a cost management plan. The cost management plan is a subset of the project management plan that defines the procedures used in gathering, estimating, and documenting all project costs; structuring a budget; and monitoring, controlling, and reporting processes. The project manager can use this tool to define more specific information, such as

- Roles, responsibilities, and authority
- Potential sources of data, units of measure, level of precision, and accuracy
- Procedures, tools, and techniques for cost estimating and budget development
- Monitoring, controls, reporting formats, and communication matrices

9

Cost Estimating

Introduction

The task of estimating cost for project activities is typically performed after the project deliverable has been broken down into its smallest components called *work package activities* and all work package activity information has been gathered. It is also beneficial for the work activities to be scheduled in a work breakdown structure so that sequencing and the connection of activities can be analyzed. It is important to define this information prior to cost estimating because specific work activity requirements and the relationship of work activity connections within a network might have an impact on cost estimates.

The project manager has the responsibility of structuring project activity costs and also monitoring, controlling, and reporting activity costs as compared to the project budget. It is important the budget is created carefully, reflecting estimates that are as realistic as possible so that actual costs have a higher probability of matching original budget estimates. The project manager has to control activities, resources, and procurements to manage costs within the estimated budget. Therefore, he needs to understand and define the processes in gathering and estimating activity costs for the other project staff who will be assisting in gathering and estimating costs for project activities. This chapter addresses potential issues the project manager faces in estimating project activity costs, such as collecting data, identifying cost constraints, and estimating tools and techniques in the quality of cost.

Collecting Cost Data

As with schedule estimating, the first things the project manager should consider in estimating project costs are details defining the collection of data for costs on project activities. It is important the project manager understand the impact of accurately estimating project activity costs because these costs can create financial problems not only at the project level, but also within the organization and with vendor/supplier and customer relationships. Depending on the size and complexity of a project, monthly cash flow within the organization is scheduled based on project estimates, and larger cost activities—if not controlled within a budget—can impact other expenditures required within the organization.

In some cases, the project manager must be aware of a project's contractual obligations that can have financial ramifications if activities are not performed as agreed upon within the contract. Not meeting the terms of the contract can also put the organization into legal situations that can present bigger financial problems it will have to deal with. Contracting human resources for specific work activity services may require payment schedules and reimbursement clauses that make estimating resources for that activity more complex, so special considerations must be taken within the budget.

Because many details can influence the estimation of project costs, this chapter focuses on understanding activity details to ensure everything within a work package has been accounted for when finalizing estimated costs of that work package activity. This requires the project manager to revisit all the information gathered and the requirements for each work activity to utilize tools and techniques to ensure the reliability and accuracy of cost information used for estimating.

Identify Cost Requirements

As the project manager begins to go through each of the work activities, review of activity requirements presents the first level of costs that need to be evaluated for estimates used in the project budget. When the project manager is reviewing the information gathered for each work activity, it is also important to note what items have costs associated with requirements and what costs are added features or additional items that may not necessarily be requirements, but instead are additional items, features, or options that might be considered if the budget allows. This review creates a list of *absolute requirement costs* and *added feature costs* and is considered depending on budget allowances.

Based on the original negotiation of the project objective with the customer, a rough budget estimate probably was identified in the original statement of work and charter that authorized the project. In most cases, if those in the original discussions understood this value to be a rough estimate, the project manager might have leeway in that the actual budget estimate can be based on real data derived from all activity requirements and development of a finalized budget. In some cases, the original estimate needs to be adhered to, either by verbal agreement or contractually. This budget value, therefore, is a project constraint the project manager has to operate within and is a major consideration when compiling estimates to create the project budget.

The project manager needs to make a fundamental decision in reviewing project activity costs and all other costs associated with conducting the project. The purpose is to identify the difference between absolute required costs and extra or added feature costs. It is generally in the best interest of both the organization and the customer that project budgets are developed with the understanding that they have the least amount of cost possible. Care must be taken in developing a budget that does not have activities defining costs outside the scope of what is really required for those activities. Activities exhibiting additional scope, sometimes called *scope creep*, artificially inflate the budget and, in most cases, are not required. This may simply be the result of improperly identifying absolute required costs versus additional feature costs and can be approved if budget, schedule, and the quality of the project deliverable permit, or can be rejected as unnecessary.

Cost Data Sources

Organizations, whether large or small, potentially have information on items that have been purchased or rented or labor rates for human resources that have been used in the past for operations and project activities. When project managers and assisting project staff are collecting information for cost estimates, this information can be found in historical data, calculated based on historical data, or acquired by soliciting cost estimates from suppliers or vendors providing resources for project activities. When the project manager determines that cost estimating data has to be gathered, there are several considerations regarding how this task is accomplished that can influence the quality cost estimates. Depending on the size and complexity of a project and the size and structure of the organization, a project manager may collect cost estimates on her own or may solicit the help of other project staff and associates who can assist in cost data collection. In any case, some important elements of data gathering have to be considered to ensure cost estimating is performed accurately and correctly:

Who will be collecting information? Individuals selected for gathering cost information on work activities need to be qualified in what cost information they will be gathering. Individuals with experience on a particular activity have an intuitive sense of the validity of cost estimates, know where to seek accurate and reliable information, and know what questions to ask to obtain accurate cost estimates. Having the correct individuals collecting data also helps expedite the data-gathering process in the interest of developing the overall project budget as quickly as possible.

What information should be collected? Individuals selected for gathering cost information also need to know where to look for information that needs to be acquired. If project deliverables have been broken down into their smallest work activity components, this is the first place individuals go to understand the scope of information required to accurately estimate costs for requirements within a particular activity. Information should be as complete and detailed as possible, estimating all costs associated with a requirement that may include special options such as features; shipping requirements; and any expediting fees, taxes, or permits that need to be purchased in association with an activity.

What sources of information are reliable? The next area of importance in collecting information is the quality and reliability of the information source. It is typically understood that historical data that has been well documented for specific work activities can be a source of reliable information. The key in using historical data is determining whether the features and parameters of a previous project activity are comparable to use as information for estimating a new project activity. Individuals collecting information should always try to use firsthand information, whether it was recorded data or gathered through interviews or surveys with individuals. Information solicited from vendors or suppliers typically is reliable and accurate because it is normally associated with a written quote or bid or is priced out of a documented price sheet. These sources are both reliable and accurate because they typically are updated and relevant, and they are specific to the activity and requirements information provided.

How much information should be collected? Individuals invest time and effort into investigating cost information and should make the best use of this time to ensure an adequate amount of information is collected. It is also important the project manager ensure that the primary deliverable has been broken down into the smallest components possible and detailed information is gathered for activity requirements so that individuals collecting data know how much information should be collected. When one is collecting cost information, it is best to gather as much information as possible and later determine what information is necessary. Discarding information is much easier than having to return to the source and collect more information. It is also important that the information be as detailed as possible because this helps ensure the cost estimate will be accurate.

Data Accuracy

For the project manager, the ultimate goal in collecting cost information from project activities is to ensure its accuracy and reliability. As discussed previously, reliability is important because it is a key element in where information was derived. The second element of information gathering is accuracy. Those tasked with the collection of information need to understand the importance of accuracy and the role it plays in cost estimating. Accuracy refers to how close an estimated value is to the actual value. The importance of understanding the impact accuracy has on estimating will be seen when the project is underway and actual costs are compared to the original estimates documented within the budget. If a project budget has been established based on these estimates, it is important the actual costs be as close to these estimates as possible; otherwise, the project runs the risk of going over budget, requiring more funding.

The project manager has two ways to look at managing a budget in the course of a project: (1) ensure estimated costs are as accurate as possible so that there are little or no variances in actual costs, thus requiring minimal cost controls; or (2) develop a budget with less accurate estimates and try to control cost overruns during the course of the project. It is clear that time well spent in gathering cost information with the highest accuracy possible is preferred in managing a budget over the course of a project life cycle.

Cost Classifications

Within organizations, structures define work activities required to manage the operation. Depending on their structure, departments within functional organizations complete specific activities required by each department for the overall organization, whereas in projectized organizations, specific activities are performed to complete a project objective. Each structure requires resources and materials to perform these activities, and they may or may not be directly related to project activities. Projects have a specific objective to create a unique deliverable, and that deliverable is broken down into specific work activities, each requiring resources and materials to be accomplished. Those activity costs are classified as *direct costs*. Projects also involve other activities that are required from other supporting departments within the organization, and they are classified as *indirect costs*.

Direct costs—These costs are associated directly with work activities required to produce project deliverables. Direct costs are typically estimated within the project budget and recorded within the project baseline. Project work activities usually require several, if not all, of the following examples of direct costs:

Labor directly associated with project work activity

Materials directly associated with project work activity

Equipment or other **resources** allocated or procured for a specific work activity

Consulting or **contracting** outsourced services required for the project

Travel expenses incurred specific to project work activity

Indirect costs—These costs include all other operations costs incurred by the organization to support the production of a project deliverable. Indirect costs are typically not associated with the project budget and are not generally included in the project baseline but, depending on the organizational structure, may be tracked as part of project costs. Most organizations have many or all of the following indirect cost components as a function of maintaining an operation and supporting project activities:

Overhead expenses—The organization has these expenses as a function of general operations and not associated with any particular project, including facilities, utilities, equipment not used by a particular project, depreciation of equipment, telephone and computer services, corporate and employee insurance, marketing and advertising, payroll taxes, and employee benefits.

Administrative costs—The organization has these expenses in managing general operations, which can include management and administration functions performing duties such as accounting, human resources, IT support, legal and contract management, and facilities and grounds support.

Indirect labor—These labor costs are associated with all operations functions conducted in and around project activities but are not directly associated with any work activity.

Indirect materials—These costs include all materials procured for use by the organization on a daily basis that were not specifically identified for project activities, or may be used on several projects without discretion. Examples include office supplies, copy paper, pencils and pens, tape, and so on.

Cost Constraints

As project managers review requirements of project activities, sometimes certain limitations, regulations, or processes present constraints in the cost estimating exercise. Those tasked with collecting cost information may run into situations in which constraints create a challenge as to the availability of accurate cost estimates. This may be due to a lack of information or a limitation or regulation imposed on a component of work activity forming a constraint. These types of challenges are typically found at one of three levels: the organizational, project, or customer level.

Organizational Level

Organizational-level constraints are typically found when management or other individuals within the organization have influence on cost estimating, such as insisting on a particular process, material, or work activity component that was not originally defined for that work activity; or insisting on a particular vendor supplier that may not provide the most optimum cost. Procurement departments sometimes impose constraints when using preferred supplier/vendor lists that limit where procurements can be made; these limitations can have an influence on cost estimations. It may be determined the organization has a specified process or procedure for a work activity that was not originally identified in the work activity requirements; such specifications impose a constraint on cost estimating. It may also be determined that those in the organization do not have the technical or intellectual capacity to produce a requirement of a work activity, and this imposes a constraint on accurate cost estimating to complete the activity.

Project Level

Constraints for cost estimating can also be found at the project level, where challenges are generally identified by the project manager and/or other project staff. The project manager may determine that the organization lacks individuals qualified to correctly gather accurate cost estimating data. Having an inadequate staff of qualified individuals will result in a smaller number of people taking a longer period of time to collect data, or the organization may contract with external resources experienced and qualified with specific cost information gathering exercises. The project manager may also find that there was not enough time to gather complete and accurate data for cost estimating, so he may be forced to use more expedient estimating methods that simply lack in accuracy.

Customer Level

The final form of cost estimating constraint usually is a result of customer requirements and/or changes. In the initial planning stages of a project objective, it is typical for there to be several conversations concerning details of deliverables and changes or extra requirements imposed. If cost estimating takes place during this phase, it is difficult for the project manager to maintain accuracy of cost estimates as a result of changing requirements. This situation creates a constraint that requires either waiting until all details have been finalized before cost estimating can commence or accepting variance within the cost estimates based on the fluctuating nature of defining a project deliverable. In most cases, more specific cost estimates are defined after the project deliverable has been defined, but in some cases, this activity is required to formulate an initial project cost estimate for general budgetary purposes within the statement of work or project charter.

Customers may also require specific processes or procedures that impose constraints on the flexibility of cost estimating for a particular work activity. For example, customers may require the use of specific suppliers or vendors for specialized procurements, which reduces the flexibility of shopping for best cost estimates. Customers also may require a detailed estimate of budget at completion that does not allow enough time to gather detailed and accurate cost estimates, thus forcing the project manager to use less accurate estimating methods. Customers may even insist on using specific cost estimating methods, which also limits the flexibility the project manager may need in cost estimating due to the availability of accurate information.

Estimating Tools and Techniques

The primary goal of the project manager is to establish accurate and relevant cost estimates for all activities and periphery items associated with the project to create the project budget. In doing so, the project manager needs to understand where accurate and reliable estimations can be derived from, consider any constraints that may influence cost estimating, and use tools and techniques designed to assist him and/or associated project staff in developing accurate cost estimates. As with schedule estimating, the project manager can use similar tools in cost estimating; they are covered later in this chapter. The project manager should always strive to get the most accurate data possible and may have sources that allow estimating tools to use absolute data. If objective and absolute cost data is not available for certain items within project activities, he may have to use other tools to formulate cost estimations.

Subject Matter Expert

A subject matter expert (SME) is someone who has experience, skills, and knowledge pertaining to a specific element within a work activity. This person can provide a rough approximation of cost if more precise cost estimating is not available. Having SMEs estimate cost typically occurs during the conceptual phase of a project in developing a statement of work or project charter where rough orders of magnitude type data would be used for initial budget estimating. It should be noted that in discussions with subject matter experts, project managers might discover other undocumented but relevant pieces of information about elements of work activity that normally may not have been considered and that may influence cost estimating.

Rough Order of Magnitude Estimating

If senior management, initial stakeholders, and project managers have been selected prior to project approval, when they need to compile a generalized rough estimate at completion for budgetary planning, it is most common to use *rough order of magnitude estimating* (ROME). Senior and functional managers and initial project stakeholders can make feasibility decisions as to general financial requirements, potential return on investment (ROI), as well as cash flow requirements of the organization to support a potential project. It is typical to solicit individuals within management or subject matter experts to acquire generalized rough approximation information that can be used for feasibility studies of projects. This form of estimating should be used only at the onset of a project when little detail has been determined and rough estimates suffice for an overall project budgetary analysis.

Analogous Cost Estimating

Analogous cost estimating uses information from previous project activities that are similar in scope, complexity, and requirements of deliverables as a comparable for estimating costs of new project activities. This approach is based on the analogy that if a previous activity had a particular cost and by comparison is similar enough to an activity in the new project, the cost documented in the previous project can be used as an estimate for the activity in the new project. Analogous cost estimating can also be less time-consuming because the project manager is simply using comparison analysis to form general estimates. It should also be noted that this form of estimating may not be as accurate as other forms of estimating because other factors of influence and specific parameter attributes may not always be comparable and may affect the accuracy of the estimation.

Example: A project manager is estimating the cost of labor required for a construction project activity. Because it may be difficult to estimate the man-hours to complete the specific activity, he reviews a project that was completed earlier in the year that has the same scope of work, materials used, and tools available. He can then determine that the current activity is similar to the same work activity in the previous project; consequently, he can use the number of man-hours recorded and labor rate associated with that activity for the estimate of the same activity on the new project.

Parametric Cost Estimating

Parametric cost estimating uses the same bases of analogous estimating from historical data of previous projects and similar activities but utilizes a statistical or scalable component to derive a relevant cost estimate. Sometimes historical data can be used to identify a component work activity that is similar to an activity on a new project but has a difference in size, shape, or feature that will not allow the previous documented cost to be used one for one in the activity of the new project. In this case, data from the historical activity component can then be scaled to match the parameters of the new activity, thus producing a cost value that is relevant to that activity.

Example: A project manager is estimating the cost of an IT installation project that she discovers is using the same equipment as used on a prior project. She discovers that two sets of equipment were used on the prior project and the activities for which she is estimating costs use five sets of the same equipment. She can then take the cost of the prior two sets of equipment and multiply it by 2 to derive an estimate for her new project, as follows:

Prior project activity: server equipment = (2) racks @ $55,000 total
New project activity: server equipment = (4) racks = 2 × $55,000 = $110,000
New project activity: server equipment cost estimate = $110,000

Three-Point Cost Estimating

Project managers may find cost information of components within work activities that would suggest not only one absolute value, but also a range of values that might represent a variety of influences that could swing the cost to a positive or negative direction. If this is the case, it should be important not to discount the information causing the influences because it might be relevant and important to note within the estimated cost. If a range of information that potentially influences cost is available, the project manager must have a tool to incorporate such a range and derive one single value of cost.

Three-point cost estimating utilizes both *optimistic* and *pessimistic* values to calculate an expected cost. Two forms of this method are the *triangular distribution* and the *beta distribution*. The beta distribution form of three-point estimating method was originally developed as part of the program evaluation and review technique (PERT).

Optimistic (C_o)—This estimate is based on data that would suggest an absolute best-case scenario condition for a work activity. This might represent the lowest possible cost of an item; the availability of a rebate situation that would decrease cost if certain conditions are met; or a special sale that, if acted upon within a certain time frame, would result in a lowered cost. Any of these conditions might be available, and the project manager might deem it necessary to include the possibility of this more optimistic value of cost.

Most likely (C_m)—This estimate is based on data that would suggest a nominal value of cost with little or no risk or constraint influence. This would represent the normal everyday cost of an item that is most likely to occur given no anticipated changes, fluctuations, or influences that would alter the normal price.

Pessimistic (C_p)—This estimate is based on data that would suggest an absolute worst-case scenario for a work activity. The pessimistic value would typically infer problems that might be associated with procuring particular items or issues related to labor costs. This might be the potential unavailability of the intended item and expediting fees that might have to be incorporated to meet schedule requirements, shipping damages that may require rework of an item at a higher cost, or the purchase of an additional item. Any number of things can add cost to an item including the influence a potential risk might incur.

Expected cost (C_e)—The expected cost is the mean of the distribution calculations (shown here) taking into consideration all three classifications of estimates: *optimistic, most likely,* and *pessimistic*. The project manager can use this value as a cost estimate outcome for a particular work activity.

Triangular Distribution: $C_e = \dfrac{(C_o + C_m + C_p)}{3}$

Beta Distribution (PERT): $C_e = \dfrac{(C_o + 4C_m + C_p)}{6}$

The beta distribution (PERT) formula is shown in the following example and illustrated in Figure 9.1 for the remaining project activity estimates. Note how the distribution works to skew the expected (C_e) value from the most likely (C_m)

in the direction that emphasizes a probability of more positive or negative effect to the original (most likely) estimate. This distribution can be valuable if information is available to include an optimistic or pessimistic influence to an estimate.

Activity A Data: Optimistic = $1,100, Most Likely = $1,250, Pessimistic = $1,500

$$C_e = \frac{(1,100 + 4(1,250) + 1,500)}{6}$$

$$C_e = \$1,267$$

Activity	Beta Distribution for Activity Cost Estimating			
	Optimistic	Most Likely	Pessimistic	Expected
A	$1,100	$1,250	$1,500	$1,267
B	$6,575	$7,640	$7,850	$7,498
C	$9,500	$10,375	$11,000	$10,333
D	$26,250	$27,300	$28,100	$27,258
E	$8,900	$9,125	$9,250	$9,108
F	$2,000	$2,160	$2,200	$2,140

Figure 9.1 Beta distribution of cost estimates

The beta distribution is similar to the triangular distribution but applies more emphasis to the *most likely* data and deemphasizes the two extremes (*optimistic* and *pessimistic*), as shown in the formula where C_m has a multiplier of 4 and is divided by a factor of 6 to normalize the magnitude of distribution. This allows the project manager to take into consideration the influence of the two extremes, but they will carry less weight in the overall calculation of the cost estimate.

Top-Down and Bottom-Up Estimating

Project cost estimates can be derived from two different perspectives relative to the project in general: (1) as seen from the top, viewing an entire project and using generalized estimations of project cost; or (2) from the bottom, rolling up specific data of each work activity into a finalized project budget. Because both forms of estimating are used in organizations for project budget planning, most project managers should elect to use a bottom-up form of estimating to capture the most amount of detail relative to activity costs.

Top-down—This type of estimating is generally less accurate; it is used at the beginning of the project when developing the overall scope statement or project charter for rough orders of magnitude estimating in general project planning within the organization. Top-down estimating is less accurate for specific activity components and, in most cases, derived from management or subject matter experts using generalized data primarily based on historical project activity packages.

Bottom-up—This type of estimating is generally more accurate because it starts at the bottom with the detailed costs of all specific components of each work activity and progressively forms the sum of all costs used in estimating the project budget. Bottom-up estimating would be the logical step after a project deliverable has been broken down into its smallest components or work activities and detailed information has been gathered regarding the requirements of each work activity. Only then would the project manager feel most comfortable preparing cost estimates for each work activity requirement. This approach also allows for information such as specific risk considerations, contingencies, and specific cost considerations for detailed items within a work activity that would not be captured in a top-down perspective.

Contingency Cost Estimating

When project managers are evaluating work activity requirements, certain risks, uncertainties, and conditions that are associated with the procurement of resources, materials, and labor must be factored into cost estimates. As you have seen with estimating schedule durations and the development of an overall project schedule, buffering considerations can be applied at the activity level for specific items as well as the overall project level to account for uncertainties; these considerations are called *contingency cost estimating*.

Contingency cost estimating, sometimes called *reserve analysis,* requires not only the analysis of the expected cost of a component within a work activity, but also the evaluation of other influences that may increase the cost of certain items. The project manager might elect to use three-point estimating to factor in the possibility of negative influences in estimating costs, but the outcome will be based on a standard distribution mean and not an absolute number used to increase the cost estimate based on a particular concern. Contingency estimating usually requires the project manager to perform a basic review or analysis of potential risk for each procurement identified; the purpose is to identify the impact and probability it might have to the overall budget and whether or not reserve funding or contingencies should be planned within the cost estimate.

The project manager should be very careful in contingency estimating because this directly increases the cost of identified items and influences the overall proposed budget. If contingency estimating can be planned into the budget and recorded within the baseline at the beginning of the project, this approach is preferred because the overall budget can be approved with all contingency plans in place. If the budget is approved without contingency planning as part of the baseline, the project manager is forced to control costs of each work activity, and regardless of identified risks, that may prove difficult based on the probability and impact a potential risk could have to the budget. It is typically better to plan contingency estimates where needed and have them be approved in the budget and recorded on the baseline because this gives the project manager the assurance that high probability and impact risks have been accounted for, and the project is likely to stay on budget and not require additional funding.

Review Questions

1. Explain the significance that data accuracy can have in cost estimating.
2. Discuss the differences in constraints at the organizational, project, and customer levels.
3. Explain the benefits of information gathered from subject matter experts.
4. What conditions make using three-point estimating justifiable?
5. Explain contingency cost estimating.

Applications Exercise for Chapters 9 and 10

Millbrae Hander Medical Center Expansion: Case Study

The Millbrae Hander Medical Center has been an inner-city outpatient medical facility for 12 years and is in need of expansion to include more examination rooms as well as the addition of inpatient rooms. Millbrae Hander, founder of the medical center, passed away 5 years after the facility was opened, and the Millbrae Hander Foundation has succeeded in managing the operations of the medical center in his absence. This medical center has become a landmark facility within the community because it not only caters to the inner-city community but also has served as an educational platform for medical interns. The medical center, currently sized at 36,000 square feet, is suffering from overcrowding in the limitation of outpatient service only and lack of examination rooms and staff preparation areas. The medical center has undergone two small remodeling endeavors to remedy this situation, but the increased demand for medical services continues to put pressure on the size of this much-needed medical facility.

City planning has authorized an expansion of 10,000 square feet to the existing facility and use of property and parking adjacent to the existing facility that the medical center currently owns. The medical center has a combination of two large donations and has secured low-interest financing to manage the remainder of the total financial responsibility. The director and staff of the medical facility are excited about the approval of this expansion and want to move forward with project planning.

An architectural firm has been identified to produce all building specifications, drawings, and documentation necessary for city permits to be drawn. A project management firm has been identified to manage the construction and move in phases for this project. Representatives from the Millbrae Hander Foundation, the executive director of the medical center, and key department staff supervisors have formed a panel of stakeholders to identify what is needed in the expansion of the medical center. A list of critical areas and a general project budget have been formulated and communicated to the architectural firm, and an overall project duration of ten months has been established for completion of the expansion project. At this point, the project manager only has financial information for the building and land at $525,000 and an initial budget of $725,000 for medical equipment and supplies. The stakeholder panel has issued an initial project budget of $1,300,000 and has communicated that only $25,000 emergency buffers are available. The stakeholder panel has also issued the requirements as to what will be included in the expansion, as listed here:

Quantity	Description	Size	Square Footage
6	Staff offices	12 × 12	864
22	Inpatient rooms	14 × 16	4,928
10	Examination rooms	12 × 12	1,440
2	Staff prep areas	various	1,264
2	Restrooms	14 × 20	560
1	Waiting area	18 × 24	432
1	Staff break area	14 × 16	224
2	Storage rooms	12 × 12	288

Apply the concepts described in this chapter to the case study:

1. Identify other necessary items that do not have assigned costs, such as architect fees, permits, parking lot, landscaping and signage, furniture, and so on.
2. Identify sources for cost information on items identified.
3. Are there any cost constraints?
4. Use at least two different cost estimating techniques in developing the cost estimates for this project.

10

Budget Development

Introduction

In most organizations, senior management has the responsibility to approve and assign oversight of activities, including projects, and therefore utilize the resources within the organization. A primary task within most organizations is that of directing the distribution of finances to manage the day-to-day operations. If an organization is going to utilize a project, that organization needs to develop an overall cost of this project so that management can decide whether the project is not only feasible, but also of value to the organization. The cost of a project is usually defined by the sum of costs related to all activities required to accomplish the project objective, and this is called the *project budget*.

In many cases, information outlining the objective of a project early in the conceptual phase may identify high-level budgetary cost values that may represent larger components of a proposed deliverable. This information is typically used so that management can have a rough order of magnitude understanding of what the overall project might cost. If a project is approved and a project manager is assigned to oversee the development of a project management plan, more specific costs related to detailed work activities need to be estimated and compiled within a budget to establish a more precise cost of the overall project. The project manager also needs a financial plan for the project to monitor and control activities; that way, she can ensure that funding the organization anticipated to be spent on work activities matches that of actual spending. The project manager can also use the budget to control spending to ensure activity costs do not exceed budgeted estimates. This chapter explores methods used in developing project budgets, along with their functionality and constraints that can assist the project manager and associated project staff who assist the project manager in developing an overall project budget.

Functions of a Budget

As you have found with developing the master schedule of work activities and how activities are connected within a network, it is also important to develop a network of costs associated with each activity to document financial information for the entire project. Creating the master schedule of work activities and creating the overall project financial plan or budget serve similar purposes for the project manager and others within the organization who require this type of information. The project manager typically uses a project budget for some or all of the following reasons:

- The budget identifies all costs associated with requirements for each project activity.
- The sum of all activity costs creates the overall project cost.
- The project manager can monitor actual costs in comparison to budgeted costs for each activity.
- The project manager can initiate controls to keep actual costs as close to budgeted costs as possible.
- The project manager can report financial status of work activity costs throughout the project life cycle.
- After the project is completed, the budget can be archived for historical data and therefore can be used when estimating budgets for future projects.

The project manager typically uses the budget for the preceding reasons, but the budget also holds a fundamentally more important role in why it's created and the importance of what having and maintaining a budget really means to the project and the organization. The following sections explore important reasons that budgets are created and used throughout an organization, including the management of work activities on a project and the overall utilization of resources within the organization.

Budget at Completion (BAC)

Developing a budget is a formidable task typically performed on most projects that have cost components associated with work activities. There are two primary reasons an organization needs to understand the financial basis associated with a project:

- Details of specific expenditures for work activity requirements
- Overall cost of the entire project for organizational budgetary planning

As the project progresses through the completion of work activities, the organization requires cash flow to cover the expenses outlined in the budget for each work activity. Within each work activity, the total cost for all requirements is outlined to accomplish the objective of that work activity; that is called the *activity budget at completion* (BAC). As the cost of each work activity is defined, the project manager can then add all the associated activity costs of the entire project that will produce a single budgetary value called the *project budget at completion*. Depending on the size and complexity of a project, each work activity may have to manage its own BAC, because some single project work activities may be in the millions of dollars and might have to be managed as single entities within the project. All projects have an overall budgeted cost, and in most cases, the BAC is used to identify the entire project cost (budget at completion).

Project Budget Baseline

When cost estimates of each activity are identified and an overall budget is established, the project manager essentially has a list of "expected" costs before the project starts. This estimated budget of expected costs is used as a baseline to compare estimated costs to actual costs as they occur. The project manager uses the budget baseline to monitor and implement controls to manage actual expenditures to match estimated costs. The estimated budget is one of the primary tools the project manager uses as a baseline throughout the course of a project life cycle to manage overall expenditures on a project. The processes for developing and using the project budget baseline are covered in more detail in Chapters 11, "Schedule and Cost Monitoring," and 12, "Schedule and Cost Control."

Manage the Triple Constraint

Another important reason for establishing a project budget and creating a budget baseline is to help control costs when other project influences force those involved to make decisions that can impact actual costs. The project manager might find that a work activity's actual cost is staying within budget, but trouble has developed with some component of the activity that may affect either the time or quality of the deliverable. Therefore, the organization might have to make decisions about staying on schedule or within the estimated scope of the deliverable that will cost more money; this is called *managing the triple constraint* (see Figure 10.1).

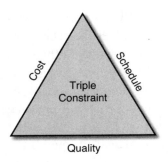

Figure 10.1 Triple constraint

The project schedule is developed based on specific *estimates of durations* for work activity that are derived from *requirements of the activity deliverable* that have an *associated cost* to produce. This forms three connecting items: time, cost, and quality associated with the production of an activity deliverable. Creating a project schedule that shows the duration and targeted work to be performed in conjunction with the budget baseline that establishes a cost to produce that work involves three parameters or pressure points (*triple constraint*) that have to be maintained to accomplish a work activity as designed. Creating the budget baseline is as important as creating the project schedule and the scope statement of work (SOW) defining work activity deliverables.

Reporting Project Status

As the project moves throughout the project life cycle, there will be periods of accountability that the project manager has to respond to; she will need to provide information regarding the status of work activities, whether the project is on schedule, and the financial status of expenditures relative to the estimated budget. The customer might also require reports at certain milestones as to the status of the work activity and the status of the schedule. Interest in reporting the financial status of a project is typically an internal requirement made by management; the purpose is to account for actual expenditures compared to the estimated budget and schedule status of the progress of work activities.

Budget Development Methods

Depending on the size and structure of an organization and whether or not that organization has a project management office (PMO), there may be requirements, processes, and procedures dictating how budgets are developed and documented. It is always best for the project manager being tasked with budget development to solicit information within the organization, PMO, or financial department as to the proper or accepted way to develop a project budget. If other projects are being managed within the organization, the project manager may seek the advice or guidance of other project managers in the development of a budget to stay consistent with other projects. Regardless of what information for project budget structuring and development is available, the project manager should be aware of some of the basic budget structures and how budgets can be developed based on available information.

Regardless of what type of budget development the project manager uses, some key components form the philosophy behind the idea of budgeting project costs and how budgets can be effective and consistent in managing the project:

- The budget aligns with all work activities identified in the work breakdown structure (WBS).
- It identifies cost estimates of all activity requirements.
- It uses a standard unit of measure across all cost estimates.
- Final project cost represents all work activity cost estimates.
- Actual costs are assessed in comparison to estimated costs.

Top-Down Budgeting

Because a project manager can develop a budget in several different ways, some organizations have more influence in this process than others, depending on the size and structure of the organization. One mode of influence an organization can have in developing a project budget is the involvement of upper management in either the creation of or involvement in creating the project budget; this is called *top-down budgeting*.

In top-down budgeting, upper management staff have either direct influence with input or create the budget entirely and it is directed to the project manager for implementation. The general idea behind top-down budgeting is that senior management have knowledge and experience with a particular subject or with past projects of a similar nature such that they can produce viable and accurate cost information that can be used in formulating a project budget. In some cases, this budget is created entirely within upper management and is given to a project manager to implement, whereas in other cases, the project manager is tasked with developing the project but will use input from senior management.

The advantages of top-down budgeting are typically found within the level of experience of senior management who are involved in creating an accurate budget. In smaller organizations where there may be a lack of specific project information for cost estimates, senior management may be the best source of accurate information to create a budget. A disadvantage of this type of budgeting occurs when senior management staff do not have the experience or background that will yield accurate cost information, but they insist on participating in budget development nonetheless. Other levels of management and project staff may not agree with the assessment of certain activity cost estimates senior management use in the budget, and this can cause conflict as to potential funding available for the project. In some cases, the specific agendas of certain managers may be in conflict with senior managers developing a budget with regard to utilization of resources, hierarchy of project importance, and scheduling and allocation of critical funding.

Cost Aggregation Method

Cost aggregation budgeting uses a method of budget structuring and information gathering that requires the project deliverable be broken down into its smallest components and information gathered on the requirements of specific work activities. Costs are estimated on each specific component of work within an activity and produce a high level of accuracy for project cost estimating. Cost estimates for each work activity are then aggregated to the next level component in the work breakdown structure and continue up until all project costs are combined into one final project cost; this is then used as the project budget, a process otherwise referred to as *bottom-up budgeting*.

Cost aggregate budgeting requires a tremendous amount of work by the project manager and other project staff. They must correctly break down the project deliverables into their smallest components and effectively identify critical information regarding each work activity so that cost estimating can be as accurate as possible. An advantage of the bottom-up estimating approach is that it is based on the smallest levels of work and therefore has the highest accuracy of cost estimating, resulting in a more accurate total project estimated budget. Another advantage is in monitoring and controlling a bottom-up budget because this also gives the project manager detailed cost information that will help in controlling expenditures and procurements for project activities. The primary disadvantage with bottom-up budgeting is the amount of time and work required to identify activity information and cost estimates to create the budget.

Time-Phased Method

Because the project manager uses the master project schedule to control work activity durations based on original estimates, controlling these estimated durations becomes important only if they are connected in a sequence where dependencies and predecessor/successor relationships require the control of durations. If the project manager also intends to use a project budget for the purpose of cost expenditure control over the project life cycle, the budget in its informal state is simply a compilation of specific activity cost estimates and needs to be associated with the same schedule of activities; this is called *time-phased budgeting*.

The philosophy behind time-phased budgeting is that the estimated costs for each work activity scheduled in the work breakdown structure are connected to that activity. Consequently, sequencing of specific activity costs can be analyzed much like activity durations for cash flow and any funding constraints that might be generated based on the sequencing of activities. The project manager also uses the time-phased budget to see future procurements and expenditures required for work activities and to plan ahead for the availability of funding. This also allows the project manager to assess potential risk and contingency plans that might have to be implemented that may affect cash flow or funding requirements.

Time-phased budgets are typically implemented when the WBS is created and the network diagram of sequenced activities has been generated. After information is gathered on each activity and a full analysis of activities is performed to confirm the correct connection and activity relationships, cost estimates can also be input to each work activity label on the network diagram for further analysis to ensure that cash flow and funding constraints are not a problem. This also allows the project manager to control and report on project activities for one single location, seeing all three elements of the triple constraint—time, cost and activity deliverables.

Analogous Budgeting

Analogous budgeting relies on historical information that can be reviewed and used in the development of a new project budget. It is important that project managers archive past project budgets so that they can be used in future projects. They may provide information such as a budget structure based on a particular project deliverable, cost estimate values, and techniques for cash flow management and contingency planning. Analogous budgeting can be used directly on new projects in two different ways depending on the needs or requirements of a new project budget:

> **Historical derivative budget**—Analogous budgeting can be used when a historical project, by comparison, has the exact same deliverable, work breakdown structure of activities, and associated cost estimates. For example, in the creation of a single family residence, it has been determined that a new project can be formulated based on the exact same home built earlier that year for another customer. The same architectural plans, materials, and contractors can be used on the new project, therefore allowing the project manager to develop a new project plan and associated budget as a *derivative of the prior budget*. It is always prudent for the project manager to verify some of the larger cost estimates to ensure they are still accurate. In most cases, however, if the previous project was completed with little time between it and the new project, most of the costs should still be accurate and can be used on the new project.
>
> **Historical information for budget**—Analogous budgeting can also be used when a new project is similar to a historical project, but either some work activity components are different or too much time has elapsed since the last project for reliable or accurate cost estimates to be used directly. In this case, the project manager simply uses the previous project only for comparison purposes to clarify certain work activity requirements or to compare prior cost estimates to new cost estimates and provide overall budgetary validation. Historical projects can be a valuable source of information for both absolute budgetary information and "lessons learned" insights.

Budget Constraints

As project managers develop project budgets, inevitably some situations will arise that create challenges in the process. In some cases, they may simply be unfortunate events that create isolated problems that a project manager has to resolve before the completion of a project budget. In other cases, they may be more systemic issues that cannot be easily solved and are considered constraints that will impose limitations or influence the outcome of a project budget. Budget constraints can take the following form:

- Upper management's involvement or influence on particular parts of a budget, resulting in inconsistent or inaccurate data being used in cost estimates

- Organization's antiquated budgeting processes and procedures that limit the project manager's ability to develop a relevant and sophisticated project budget
- The project manager's lack of knowledge and tools for developing an adequate budget
- Time restrictions that do not allow for the proper development of a project budget
- Lack of funding available to cover project budget requirements
- Inadequate cost estimate data for developing a project budget
- Lack of project management tools required to properly structure work activities and a project budget

As project managers become more familiar with the role that they play within the structure of an organization and the requirement for project management tools and knowledge to properly structure project budgets, they may address the impact constraints can have on developing a project budget at the onset of a project to help mitigate or eliminate fundamental budget constraints. If project managers are aware of the impact certain fundamental budget constraints can have, they may be able to make changes, purchase software, or present the case to upper management in hopes of resolving constraining issues before they create problems in budget planning.

Funding Limit Reconciliation

As the project manager develops a proposed budget, it is useful for her to consult appropriate staff within the organization's financial department regarding the expected expenditures throughout the project life cycle. This is probably a necessary part of approving a project budget, but also is helpful before that time because the project manager may need to consider cash flow at various points within the project or spending limitations for certain work activity requirements in the overall budget. Some work activities might have a budgetary limit that needs to be set based on the commitment of funds available for that activity or that will be available within a particular time frame. The project manager must be aware of any budgetary limitations that have been imposed on work activities so that in the course of managing work activities she can ensure the actual cost does not exceed the approved budgeted cost; this process is called *funding limit reconciliation.* It is incumbent on the project manager to be aware of these types of limitations so she can take actions to avoid overspending of budget-limited items on work activities.

Budget Contingency Planning

In developing a project budget, the project manager uses all the cost estimates as a basis for establishing the budget. And as you have seen, cost estimates can be extremely accurate and relevant or can be very inaccurate and, in some cases, rough order of magnitude values. Because it is the goal of the project manager not only to develop the budget but also to understand the overall completion cost of a project, the budget is used in monitoring and controlling actual expenditures in hopes that the actual cost of a project is as close as possible to the estimated budget. In a best-case scenario for budget planning, the project manager would be delighted if all work activities had extremely accurate cost estimates and all activities had a relatively low probability of any risk events occurring. Because this is rarely the case, the project manager needs to evaluate work activities for possible risk and the need for extra funding to resolve issues. The project manager can choose to perform this analysis at the activity level, where contingency planning can be integrated into the actual work activity cost estimate. As this is typically the best way to design contingency planning into a project, the project manager might also elect to design contingency planning at the budget level.

If the project manager elects to use budget contingency planning, these funds are allocated to the overall project budget for use in covering unforeseen problems throughout the project life cycle. Some organizations require specific accountability of contingency planning within the actual budget. In other cases, however, organizations require only absolute and relevant cost estimates be reflected in the project budget and any contingency planning be held in reserves outside the budget but allocated to the project. In this case, the project manager has a defined budget, and all cost estimates are treated as "expected costs." The project manager oversees expenditures to stay on budget and will need to tap the external contingency fund only in the case of a risk event.

Cost of Quality

Because the project manager monitors and controls project activities, this typically points to managing three elements that all projects encounter: the triple constraint. As mentioned previously, the term *triple constraint* refers to three parameters involved in managing project activities: time, cost, and quality of the deliverable. This book is focused on factors that influence activity durations and project schedules as well as factors that can influence cost estimates in the development of a project budget, but there is an added cost that is difficult to quantify. That cost is related to the third element of the triple constraint: the cost of quality.

In most manufacturing environments, the cost of quality refers to the expenditure of finances to ensure quality measures, such as having a quality assurance department and program with quality engineers and inspectors throughout the production floor to ensure quality is maintained at a predetermined level. The cost of quality in project management refers to the third leg of the triple constraint and what effect the lack of quality will have on time and cost within the project life cycle. The cost of quality within a project life cycle relates to any adverse effect to time, cost, or quality of the project deliverable that results in reduced performance of the project.

Procurements

One area within an organization that can have a profound effect on the triple constraint of a project is procurements. All project activities throughout the project life cycle, at some point, require some form of procurement, and the management of procurements can have an impact on project performance in all three areas of the triple constraint. For example, items might be purchased but not shipped in a timely manner; the schedule suffers in the cost of quality, resulting in added cost for expediting. If incorrect items were purchased, that affects the quality of the product deliverable, but the schedule also suffers because the correct items must be purchased and money might have been wasted on the original purchase of items that cannot be returned. The project manager's responsibility is to ensure procurements are managed correctly to reduce or eliminate the potential cost of quality in maintaining project performance.

Outsource Contracting

Another area that can impact the triple constraint is the requirement of outsourced contracting and what effects outsourcing can have on project performance. If one studies estimates of resources, outsource contracting of human resources might be an unavoidable consideration; therefore, care must be taken in the contracting of external resources to protect project performance and not incur the cost of quality. Outsourcing may include the contracting of expert human resources to perform a specific activity or may be the rental of equipment. In either case, if the resource is not selected correctly and due diligence not performed, that can present an influence to project performance and is a factor adding to the cost of quality.

Another consideration regarding outsourced contracting is the various types of contracts available to be used in negotiating and documenting a legal or binding agreement. Because most contracts are used to form a balance between negotiated sides for services rendered and compensation, if contracts are not negotiated correctly, this can present a tremendous risk for the project and a potential cost driver adding to the cost of quality.

Make-or-Buy Analysis

Because most organizations try to provide materials and resources internally for project activities, sometimes a decision has to be made to have something created internally, thus requiring resources, or to purchase the item externally at a higher cost. When these decisions have to be made, the project manager should insist on an analysis as to the pros and cons of internally created items versus externally purchased items and the potential impact this decision can have on the utilization of resources and available funds for higher-cost externally purchased items.

Care must be taken to carefully analyze the availability of internal resources as well as the skill set and level of quality in creating something internally because this approach is usually preferred but runs the risk of reduced quality or extended time, both adding to the cost of quality. Likewise, purchasing an item externally may result in a higher-quality item but will be at a higher cost, again affecting the overall performance of the project and adding to the cost of quality. Both scenarios must be analyzed as to their effect on the triple constraint of the project and the overall impact of project performance as the primary consideration.

Review Questions

1. Discuss what a budget baseline is used for.
2. What is meant by managing the triple constraint?
3. Explain how the cost aggregation method works.
4. Discuss what is meant by budget contingency planning.
5. Explain the pros and cons of outsource contracting.

Applications Exercise

Millbrae Hander Medical Center: Case Study (Chapter 9)

Apply the concepts described in this chapter to the case study:

1. Choose a budget development method and explain your selection.
2. Establish a project budget baseline.
3. Is budget contingency planning necessary?

Part 4

Project Monitoring and Control

Chapter 11   Schedule and Cost Monitoring . . . 217

Chapter 12   Schedule and Cost Control 241

During the initial phase of project development, a tremendous amount of work and effort is expended in the collection of data, development of a master schedule of all project activities, and creation of a proposed budget. Although it is the project manager's role to oversee or participate in these activities, the project manager's primary role during the course of the project life cycle is in monitoring and reporting on work activity and controlling work activity to stay on schedule and budget.

The project manager therefore needs comprehensive tools and techniques to understand how to develop monitoring of activities and how to implement controls correctly to adjust work activities to stay on schedule or budget. This requires understanding organizational processes and environmental factors that influence how, when, and why controls can be imposed and for what expected outcome. The chapters in Part 4 explore tools and techniques the project manager can use in developing monitoring and controls for project work activities.

11

Schedule and Cost Monitoring

Introduction

Depending on the size and complexity of a project, it can take on characteristics that define larger components of the project management plan. They include the development of the project schedule and budget during the initiating and planning processes versus the amount of time spent during the execution process on actual work activities. For some projects, the initiating and planning processes may take several months or years, whereas the overall time span for work activities in the execution process is very short. In other cases, the initiating and planning processes may be relatively short, but the time span for each work activity is drawn out over several years because of exhaustive and labor-intensive work required.

Regardless of how many work activities and how long the execution process takes, the project manager has the primary responsibility for monitoring and reporting on work activities and implementing controls to keep work activity on schedule and on budget. Even considering how much work was required in information gathering and developing the project schedule and budget, monitoring and controlling the project are ultimately the project manager's most important roles.

To prepare for work activities to commence at the beginning of the execution phase, the project manager has to develop monitoring systems. He first needs to consider five fundamental characteristics of project activity monitoring:

- Why monitoring of work activities is important
- What activities to monitor
- What tools and techniques should be used to create monitoring systems for work activities
- What information should be expected
- How to use information gathered from monitoring work activities

This chapter explores why monitoring is important and what it means for the project manager as well as the organization to have monitoring systems in place for work activities conducted on a project. It also presents several of the more common tools and techniques used for monitoring and troubleshooting and includes typical results illustrating what is required from monitoring systems. These tools and techniques can be used on very small projects to very large and complex projects because they are fundamental in nature; therefore, they can be easily implemented for use at any level of work activity within a project.

During the initiating and planning process, the project manager and possibly project staff expend a great deal of time breaking down a project deliverable to understand its smallest components, gather as much detailed information as possible for each work activity, and develop a master schedule and proposed budget of all work activities. This work creates the foundation for the project and defines how a deliverable will be completed to accomplish a project objective. It would be incomprehensible to think that, given all the work performed to define and develop a project master plan and budget, in the course of executing the project activities, there was little or no oversight as to whether the activities were being performed correctly, on schedule, and within the budget allocated. With most organizations, organizing work activities into projects to complete an objective would not be very successful if monitoring and controlling work activities were not included in the overall project process.

Integrated Monitoring

When the project manager considers how to monitor project work activities, the first level of assessment relates to the size and complexity of the project and what type of monitoring is really required. In very small projects where there are only a few work activities and a small handful of resources, setting up a monitoring system can be as simple as using some basic observation and data recording tools, which are covered later in this chapter. In much larger and more complex projects, much more sophisticated and elaborate forms of monitoring must be implemented; they also are covered later in this chapter. In both cases, the project manager needs to develop a system to manage information collected from work activities or observations so that he can assess project performance; this is called a *project monitoring information system*.

Depending on the size and structure of an organization, the project manager may choose to use a majority of the monitoring information system components independently within the organization and simply report status.

Example: On a construction project, the project manager may monitor work activity progress and record updates on something as simple as a Microsoft Excel spreadsheet. A work breakdown structure (WBS) might even be documented on a similar spreadsheet, and between these two spreadsheets, the project manager can run the project and report status. In this example, the project manager can independently manage monitoring functions and require little, if any, organizational assets as resources for monitoring processes.

In other cases, projects performed within the organization may use human resources that require scheduling for utilization, equipment and facilities, and materials procurements that must be coordinated. In these cases, monitoring these activities may require integrating these monitoring functions within the organization. This is called an *integrated monitoring information system.*

Example: Consider a project to develop a piece of telecom equipment for a customer. This project utilizes several types of human resources with skill sets required from several different departments. It also requires an exhaustive bill of materials that includes monitoring procurements for critical components, scheduling facilities such as laboratory space for development of a prototype, and scheduling manufacturing resources to perform pilot production. All these processes require connections throughout the organization, including the network, to properly and successfully manage monitoring functions. This is when information from monitoring requires the integration of tools used in the organization to collect, store, analyze, and communicate information within the organization.

Project Monitoring Information System

For all projects, regardless of how large or small, work activities must be carried out to complete the deliverable required in the project objective. After gathering information for each work activity and developing estimates for individual activity durations and associated cost with each activity, the project manager needs to create a system of monitoring work activity to ensure actual work activity durations and costs are being performed for the project plan. To develop a monitoring information system, the project manager must first look at the reasons monitoring has to be performed:

- **Why monitoring of work activities is important**—First and foremost, the project manager must have a clear understanding why monitoring work activities is even required. During the initial conceptual phase of a project, general agreements, assumptions, and expectations are outlined in the form of a project objective. This objective may be further defined in a specific project schedule and budget on which contractual agreements can be based. The expectation, then, is that the agreed-upon deliverables will be accomplished at an expected quality level and at the expected completion date for the agreed-upon price.

It is now the project manager's responsibility to ensure the organization holds true to its commitment in what it is delivering, the completion date, and the agreed-upon price. This goal typically is realized only if the project manager has properly identified and documented all work activity requirements, and specific schedule and budget estimates have also been documented to establish the project plan. The project manager now has to monitor and control each work activity to ensure the deliverable is being produced per the project plan.

- **What to monitor**—Because the project plan typically has an outline of all specific work activities, and it is important that each work activity be performed per the project plan, the project manager needs to monitor all work activities. He also needs to ensure that monitoring each work activity includes the actual quality of work being performed, the duration of each activity, and the overall actual cost of each activity. In addition to these three primary components of work activities, other items that should be monitored can include the scope of work, human resource performance, potential risks, and information to derive activity value and stakeholder involvement.

- **What tools and techniques are used to create monitoring systems**—As mentioned previously, the project manager, depending on the type of project, can use a variety of tools and techniques to monitor systems. They can be simple in form and implementation, or can be more complex and need to be more integrated within the organization. Specific tools and techniques used for various types of monitoring are covered in great detail in the next section, "Monitoring and Analysis Tools."

- **How to use information gathered from monitoring work activities**—As each work activity is performed, and monitoring systems yield information for each work activity, the next question is what to do with the information that has been gathered. It is the project manager's responsibility to ensure information is being documented properly and is being effectively analyzed and communicated to those interested in the outcome of project activities. The primary use of information gathered is in the comparison of actual data against the project plan's estimated data to ensure quality and completeness of work, activity duration, and compliance of activity costs with the original project plan. The project manager typically arranges this information in a form that can be communicated in status updates to others in the organization who require this information. In some cases, the customer may require certain status information from project activities. The project manager's role is to effectively and accurately collect project activity data, record and analyze data, and communicate project activity status.

Monitoring and Analysis Tools

The task of monitoring project activities can be broken down into two primary functions: *gathering information* on project activities and *analyzing information* to determine the status of each component of work activity. To perform these two functions, the project manager refers to the project plan or WBS to identify the specific requirements of each work activity. Understanding this information is important because it defines the scope of what work will be accomplished for each activity. The master schedule also has information regarding the estimated time duration required to complete each work activity, and the budget has an estimated cost of everything required to accomplish each work activity. This forms the basis of information the project manager uses in comparing against actual accomplishments and procurements for each work activity. The following sections identify some of the commonly used project management tools and techniques for both gathering information and analyzing information to determine project status.

Information Gathering Tools

The project manager first sets out to establish various ways to derive information for what's actually happening on project activities. This task can take several forms depending on what's available to the project manager through the organization and with technology, but it can be in the form of simply observing project activities to derive information, attending meetings where activity updates are being reported, and soliciting information from monitoring systems that have been put in place to derive status of work activity. As with all information-gathering exercises, the integrity of the information and its reliability are of utmost importance, so this is a good reason the project manager should utilize multiple forms of information gathering.

Status Meetings

Meetings within the organization serve two primary functions: to report information to other members and to derive or solicit information from meeting attendees for a project's requirements. The project manager can call project status meetings at regular intervals to derive information directly relating to the status of a work activity. This is an excellent form of information gathering provided those offering the information have first-hand knowledge of a specific work activity so that it is accurate and reliable. Information that should be gathered or discussed within project status meetings include

- Progress of work activity accomplished since last update
- Status of current work activity relative to activity schedule
- Reporting of actual costs from procurements for work activity
- Discussion of problems or risks that have occurred or might be imminent

Subject Matter Experts

The project manager can also solicit specific work activity information from those skilled and experienced with specific knowledge of work activities; they are called subject matter experts (SMEs). These individuals are typically tasked with performing the actual work of the activity or those directly overseeing the individuals performing the work activity. Individuals actually performing work activity have first-hand information regarding the progress and any problems related to a specific activity; this information is usually considered accurate and reliable. This may be the case in construction projects, for example, where project managers can gain first-hand information directly on the job site from those performing work activities. In other cases, the project manager may solicit an informal one-on-one meeting with an individual who might not be able to attend a status meeting but can offer accurate and reliable first-hand information.

Check Charts

Another form of basic information gathering is the use of *check charts*. They list project activities and provide workers a device to check off completion points and record duration and, in some cases, cost elements of specific work activity. This type of data can be recorded on a piece of paper within a work environment or can be in the form of a spreadsheet that is available in the work area for updates. On regular intervals, the project manager collects these check charts or pulls up the spreadsheets within an organization's network to record status of work activities, schedule, and any cost information that might be included. Using check charts is a way to collect direct activity information without requiring scheduling meetings. The use of check charts is also more successful on simple work activities, where updates can be easily documented. Check charts are a good source of project status information for simple work activities, as shown in Figure 11.1.

Work Activity Name:		Date:	Staff Name:	
Activity Tasks	Worker Name	Expected Time	Actual Time	Rework required
Inspect Materials	Joe	90 min	67 min	
Prepare Tools	Sam, Joe	60 min	35 min	
Get Specifications	Charlie	30 min	22 min	
Rough Cut Materials	Sam, Joe	240 min	267 min	Broke blade, get new
Clean Edges	Sam, Joe	120 min	94 min	
Deliver Materials	Charlie	30 min	20 min	
Clean Area	Sam, Joe	60 min	55 min	
Return Tools	Sam, Joe	30 min	21 min	
Report Completion	Charlie	10 min	10 min	

Figure 11.1 Project activity check chart

Information Analysis Tools

After the project manager implements monitoring tools and techniques to gather data on work activities, the next step is to analyze what the information really says. It's important to note that there is more to monitoring than just gathering data and reporting status. The project manager is selected not only to oversee all the activities required within the project, but also to "manage" activities to compliance of the project plan. If the project manager has effectively communicated the expectation of a work activity to those performing the functions of that activity and has developed a baseline of estimates for schedule duration and cost, he must "manage" work for that activity to hit those estimated values. This task requires gathering information about the status of the work activity and analyzing what that information reveals about it. There are several ways to analyze information produced from a work activity depending on how the information will be analyzed and/or compared to other parameters that would suggest the status of work, schedule, or cost.

Project S-Curve Analysis

One basic analysis tool takes information regarding one parameter as it relates to a second parameter within either the work activity itself or the rest of the project as a whole. The *project S-curve analysis* can take information on parameters such as cost as a function of time or work activity progress and time and, using a simple grid, display an analysis. A simple S-curve analysis is used to compare actual performance to estimated performance. In the example shown in Table 11.1, data points that have been gathered show recorded cost per function along with time duration in weeks; a corresponding S-curve in Figure 11.2 shows the comparison of actual cost to project budgeted cost.

Table 11.1 Budgeted versus Actual Activity Cost over Time

	Project Cost Budget							
	Duration (Weeks)							
Activity	2	4	6	8	10	12	14	16
Design	4	8	6					
Development		2	15	21	10	4		
Test				14	12	6	4	2
Installation						2	7	3
Total Cost	4	10	21	35	22	12	11	5
Cum. Total	4	14	35	70	92	104	115	120
Actual Costs	3	10	31	74	99			

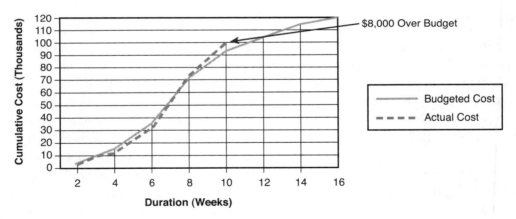

Figure 11.2 Project S-curve analysis

Milestone Analysis

Another form of analysis is the assessment of project status against major stop points or evaluation points designed into the project plan; these points are called *milestones*. Milestones can be designed into a project plan or WBS as major stop points or evaluation points of project status. In some cases, they may be regulatory inspection points that have to be designed in; they could also be engineering design review stop points where no further action can be continued unless a consensus is reached as to what has been completed. In other cases, the project manager may simply put a stopping point at the completion of major components of the project to identify significant accomplishments to the customer and/or upper management and to assess project performance for status reporting. An example of milestone status points using diamond-shaped icons is shown in the Microsoft Project illustration in Figure 11.3.

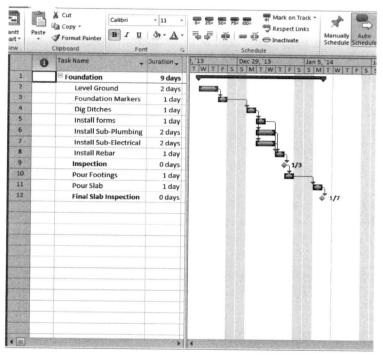

Figure 11.3 Milestone analysis in a WBS

Control Charts

Another simple form of analysis that can graphically illustrate the performance of cost or schedule is the use of *control charts*. They indicate variances of actual performance relative to a reference point. A chart is set up with two axes that indicate two parameters that are being evaluated where the zero point on the chart represents the estimate or baseline, and the plotted data represents actual performance relative to a zero point baseline. Control charts are typically used to identify a trend of information, indicating that a parameter is progressively increasing in variance in one particular direction. It is important to note that project managers have to evaluate project performance not as a function of gaining the best performance possible, but performance relative to the expected baseline or original estimate.

Control charts can be used in *trend analysis*. Where data reveals a trend going in a negative direction, this would obviously indicate poor performance is increasing, and controls need to be implemented to improve performance. Control charts can also indicate abnormal improvement in performance that may require further investigation. For instance, if a control chart is being used to track schedule and cost performance, and both indicate an abnormally high performance status, further investigation might indicate that the quality of delivered work has decreased and thus the reason that cost and schedule have improved. Control charts are excellent for revealing abnormal performance in either positive or negative directions relative to expected performance. Figure 11.4 illustrates the basic construction of a control chart given the center line is the average or mean performance ($\bar{x}$). The actual control component of this chart incorporates an upper control limit (UCL) and lower control limit (LCL) that represent three standard deviations ($\bar{x} \pm 3\sigma$) from the mean in both positive and negative directions, respectively. In addition to control limits, it is best to establish the actual specified limits that indicate a failure or out-of-spec condition; these are included on the chart designated as the upper specification limit (USL) and lower specification limit (LSL), as shown in Figure 11.4.

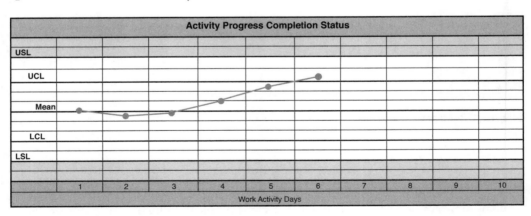

Figure 11.4 Control chart analysis of project schedule

In the case shown in Figure 11.4, using this type of monitoring tool, you can see that the status during day 6 indicates the activity is falling behind schedule. The benefit of a control chart is that the project manager can use both control limits and specification limits to determine the magnitude of the problem and how much and what kinds of controls are appropriate. This monitoring tool allows direct activity feedback to be translated into control requirements.

Create Baseline

In the course of analyzing project activities for performance, actual performance is analyzed in comparison to information gathered and documented in the project plan. This information needs to include all the work package requirements in a time-phased sequence, schedule durations for each work activity, and cost estimates for everything required within each work activity. This information can be used to form the criteria for comparison called the *baseline*.

In performing monitoring analysis, the project manager requires, at a minimum, the three criteria for performance grading as well as controlling the triple constraint: quality of the deliverable, schedule/time, and cost. These three elements are the primary information gathered on each work activity and are used at the beginning of the project to create the baseline criteria. It is important that this information is used at the *beginning* of the project to form the baseline, because it is derived from all estimates and information gathered before project activities commence. This information also represents the expectations agreed upon between the organization and customer. After the project has begun, actual work activity has started, and procurements have begun, this data cannot be used in the formation of a baseline because it is now actual data and will skew any performance comparison.

If project software such as Microsoft Project is used after the creation of the WBS, predecessor relationships have been established, and cost estimates have been included, the project manager can establish a baseline simply and easily by clicking a button, as shown in Figure 11.5.

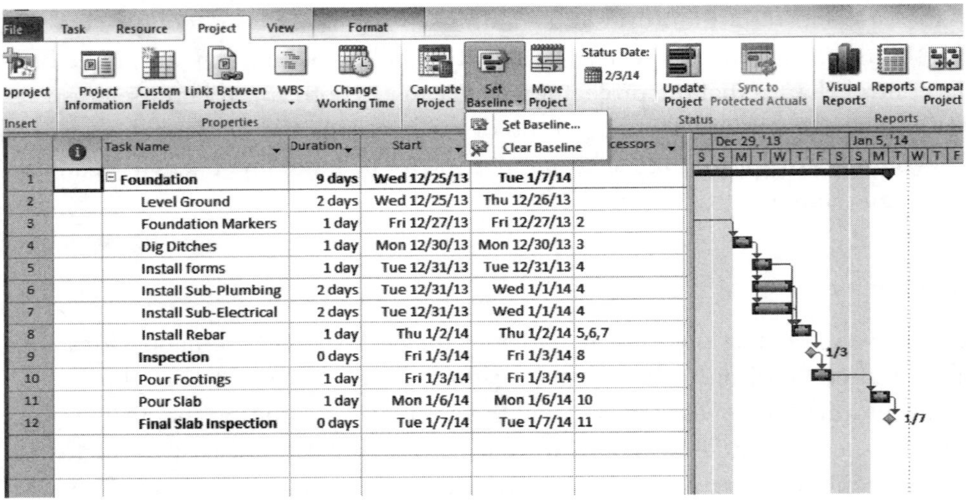

Figure 11.5 Setting the baseline in Microsoft Project

Tracking Gantt

Another tool that the project manager can use to track performance of work activities in a time-phased structure is a Tracking Gantt chart. This tool, as shown in Figure 11.6 using Microsoft Project, allows the user to view work activities on the left side of the screen. As updates of project status are included, it shows percentage of completion in reference to time. The Tracking Gantt chart is great for comparing actual performance to plan performance to help communicate, graphically, to project staff the outcome of performance analysis as a function of completed work activity. Figure 11.6 shows a WBS of activities for a construction project; the pull-down menu on the left shows how to turn on the Tracking Gantt function.

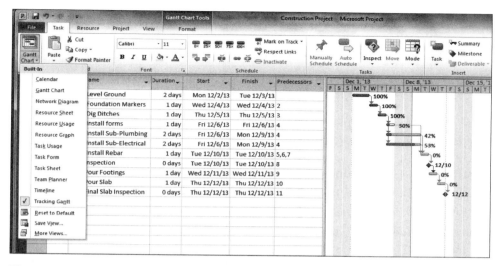

Figure 11.6 MS Project Tracking Gantt chart

Earned Value Analysis

One of the more popular forms of performance analysis used in project management today is the method of *earned value analysis* (EVA), which is more commonly referred to as *earned value management* (EVM). Most tools easily incorporate parameters such as schedule/time and cost for comparison of actual against estimates, but most typically do not easily incorporate a quantifiable analysis of work performance. The primary feature associated with the use of earned value is the ability to incorporate the progress of work activity along with schedule and cost performance. To understand how to use earned value for analysis in monitoring work activity of performance, we start with the basic terminology and associated formulas:

Planned value (PV)—Sometimes referred to as *budgeted cost of work scheduled* (BCWS), planned value is the budgeted cost for work scheduled for a particular work activity within a specified time frame. This represents the planned budget value at a given point in time during the activity. This can define one specific work activity or be the cumulative sum of all costs for a project.

PV = (Activity Total Budget) × (Scheduled % of Completion)

Example: For a single activity scheduled to be at 80% of completion:

PV = $9630 × 0.80 = $7704

Earned value (EV)—Sometimes referred to as *budgeted cost of work performed* (BCWP), earned value is the amount of actual work completed versus the budgeted cost of the same work over a specified time frame.

EV = (Activity Total Budget) × (Actual % of completion)

Example: For a single activity scheduled to be at 80% completion but *actually* at 65% completion:

EV = $9630 × 0.65 = $6260

Actual cost (AC)—Sometimes referred to as *actual cost of work performed* (ACWP), actual cost represents the actual expenditure within a specified time frame.

Cost variance (CV)—This value represents the cost variance in reference to the budget. This value is derived by subtracting actual cost from earned value. A negative value of cost variance indicates the activity or project is over budget.

CV = EV − AC

Schedule variance (SV)—This value represents a variance in schedule. This value is derived by subtracting the planned value from the earned value. A negative value in schedule variance indicates the activity or project is behind schedule.

SV = EV − PV

Cost performance index (CPI)—This value represents a ratio of earned value to actual cost. As a ratio, if the actual cost matches the earned value, the ratio value equals 1 (one), indicating the actual cost was on budget. If CPI is less than 1, the activity or project is over budget. If CPI is greater than 1, the activity or project is under budget.

$$CPI = \frac{EV}{AC}$$

Schedule performance index (SPI)—This value represents a ratio of earned value to planned value. As a ratio, if the planned value matches the earned value, the ratio value equals 1, indicating the activity is on schedule. If SPI is less than 1, the activity or project is behind schedule. If SPI is greater than 1, the activity or project is ahead of schedule.

$$SPI = \frac{EV}{PV}$$

Budget at completion (BAC)—This value represents the sum total of all estimated activity costs to form the total budgeted project cost.

Estimate to completion (ETC)—This value represents the estimate of remaining costs required to complete a project. This value is used to define the unfinished portion of a project, or from a particular point in the project to completion. ETC can be stated in the following equations:

$$ETC = EAC - AC$$

$$ETC = \frac{(BAC - EV)}{CPI}$$

Estimate at completion (EAC)—This value represents the sum total of all actual costs to date plus the estimates to complete the project. This value can be derived for various project conditions as stated in the following equations:

$$EAC = AC + ETC$$

$$EAC = AC + BAC - EV \text{ (no expected BAC variances)}$$

$$EAC = \frac{(BAC)}{CPI} \text{ (variances will continue at current CPI)}$$

Variance at completion (VAC)—This value represents the total variance in budget at the completion of the project. This value can be derived by subtracting the budget at completion from the estimate at completion.

$$VAC = BAC - EAC$$

You can now utilize data from a sample project to see how to use earned value in analyzing information derived from project activities.

Example: Initial Project Baseline Information (Schedule and Budget) Tables 11.2 through 11.4 illustrate typical project information that can be used in earned value analysis calculations to derive project performance.

Table 11.2 Project Scheduled Durations and Predecessors

Project Starting Durations and Budget			
Task	Activity Duration (Days)	Predecessor	Budget
A	3	--	$3,620
B	6	--	$8,975
C	3	A	$12,150
D	2	B	$5,160
E	3	B	$5,630
F	2	C	$2,250

Table 11.3 Project Duration in Days

Project Duration in Days (Status at end of day seven)										
Task	1	2	3	4	5	6	7	8	9	10
A										
B										
C										
D										
E										
F										

Table 11.4 Earned Value Calculations

Task	Total Activity Budget	Scheduled Completion	PV (@ Day 7)	Actual Completion (@ Day 7)	EV (BCWP)	AC (ACWP)	CV (EV- AC)	SV (EV - PV)	CPI (EV/ AC)	SPI (EV/ PV)
A	$3,620	100%	$3,620	100%	$3,620	$3,530	$90	$0	1.03	1.00
B	$8,975	100%	$8,975	100%	$8,975	$8,740	$235	$0	1.03	1.00
C	$12,150	100%	$12,150	100%	$12,150	$15,050	($2,900)	$0	0.81	1.00
D	$5,160	50%	$2,580	60%	$3,096	$3,380	($284)	$516	0.92	1.20
E	$6,360	33%	$2,099	15%	$954	$995	($41)	($1,145)	0.96	0.45
F	$2,250	0%	$0	0%	$0	$0	$0	$0	0.00	0.00
	$38,515		$29,424		$28,795	$31,695				

Based on the earned value calculation in Table 11.4, you can see that activities C, D, and E are over budget and activity E is behind schedule.

Troubleshooting Tools

As the project manager collects data on the performance of work activities, there will no doubt be occasions that the outcome of analysis reveals problems or trends that are occurring. In that case, the project manager needs to determine the cause before implementing corrective actions or controls. It is important that the root cause of a problem be determined before actions are taken so as not to overlook the actual root of a problem and implement actions on components of work activity that do not solve the actual problem. In some cases, incorrectly implementing actions or controls in the hopes of solving a problem may actually create more problems. It is important the project manager determine the actual root cause of a problem so that efforts, actions, or controls are directed at the root cause for an efficient and effective resolution.

The project manager uses the analysis tools as described in this chapter to uncover trends or potential problems but also needs analysis tools and techniques for troubleshooting to help pinpoint the root cause of a problem. The following sections describe some of the fundamental tools used in project management to utilize information gathered and analyzed from work activities to perform root cause analysis; the purpose is to uncover an actual problem or problems that affect the quality of work, schedule, or impact to budget within a work activity.

Root Cause Analysis

The first tool used to analyze information produced from work activity monitoring is called *root cause analysis*. This simple technique starts from the indication, produced by analyzing monitored information, that suggests a problem or undesired trend is present. The second component of this technique requires individuals knowledgeable in the work activity to identify all possible scenarios or "root causes" that could produce the problem. This may include one single root cause, multiple root causes, or the presence of a risk event that has occurred.

Through expert opinion, review of conditions, and further testing, the primary root cause can be determined and action can be taken to accurately address the issue. Root cause analysis is typically the most common tool used in the first pass of troubleshooting, and it has the advantage of being simple and easy to implement, producing surprisingly accurate results. A disadvantage of this technique is in the lack of knowledgeable individuals who could identify root cause scenarios; another issue could be that the work activity is too complex, requiring more sophisticated troubleshooting techniques.

Fault Tree Analysis

Fault tree analysis (FTA) is another relatively simple tool to understand and implement. It is designed to narrow down problem event scenarios through the use of an AND/OR gate fault tree to determine a root cause. This technique requires the project manager to solicit specific information from individuals knowledgeable of the work activity, review the information gathered, analyze the results that suggest a problem is occurring, and then use this information to develop the fault tree. The advantage of using the fault tree analysis is not only identifying possible problem scenarios, but also identifying the relationships that scenarios can have in creating problems. Another advantage is being able to use the fault tree in a meeting with subject matter experts participating; they can brainstorm various scenarios that might not have been otherwise considered. Figure 11.7 shows how a fault tree can be constructed.

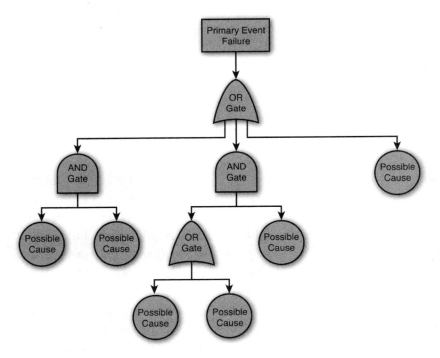

Figure 11.7 Fault tree analysis

Monitoring Results

The final step in the process of monitoring work activities is to determine how to use the information that was gathered and analyzed and who will need this information. Because projects utilize human and other organizational resources, the project manager must understand the importance of communicating project status to stakeholders and others requiring this information. If a particular work activity is falling behind schedule, or issues with procurements may cause problems with the schedule or budget, others affected by these problems need to know their magnitude and whether or not there may be a resolution. The underlying theme is that projects produce information, and it's the project manager's role to gather this information, analyze it, and determine what actions need to be performed as a result.

It may be determined that other actions are required, depending on the outcome of the analysis of work activity monitoring that may generate corrective actions, change requests, or the implementation of controls to adjust the performance of work activities. It may also be determined that other departments need to implement changes in schedules to adjust for problems on work activities within the project. The communication of information and status is of utmost importance, and the project manager must ensure that project activity monitory maintains a steady flow of information that can be analyzed for project performance.

Work Performance Reports

The first order of business the project manager has upon collecting data from monitoring work activities is to prepare work performance reports. The reports apprise stakeholders and other managers as to work activity progress and performance. Depending on the size and complexity of a project and the organizational structure, reporting of work activity progress and performance may take different forms and may be required based on milestones or other situations of reporting project status. In most cases, work performance reports communicated to the appropriate individuals generate discussions, decisions, and actions based on what the information reveals.

On larger projects, the project manager may have assistants and staff who perform analysis and take actions to mitigate or eliminate problems or risks internal to the project. Status is reported to stakeholders or upper management on the general condition of the project as a whole. In other project scenarios, the project manager may communicate the outcome of information and basic analysis to others within the organization to escalate a problem for help outside the project. This can be in the form of reports, memos, or change requests that might be required to resolve a problem situation. The project manager then looks to others within the organization for help in troubleshooting and problem resolution.

With any project scenario, the collection and analysis of data and the reporting of project activity status are of utmost importance not only because these reports provide general communication of work activity progress, but also because they inform appropriate individuals of the need to resolve problems correctly and efficiently. Performance reports can be in the form of handwritten documents or memos, or spreadsheet or text documents outlining general status information. They can be hand-delivered and/or emailed in soft copy to other project staff, stakeholders, or individuals requiring this information. Project activity status is also typically conveyed in meetings where project staff and others requiring this information can review and discuss details. These types of meetings can also facilitate off-site conference calls or videoconferencing where other project staff, management, or individuals can be informed of work activity progress and can discuss project issues. The project manager can utilize electronic whiteboards and other tools that help convey work activity details or in the discussion and analysis of information regarding problems with a particular work activity. It is important the project manager understand basic communication concepts and forms of communication to effectively and efficiently convey project information and status to individuals and stakeholders who have interest or responsibility on the project.

New Risk Assessment

Another result of project activity monitoring is in the discovery of new risks and the additional challenge that a new risk will present to project scheduling and budget control. As you have seen in this chapter, monitoring project activities reveals anomalies, problems, or trends based on information of project activities that have already occurred. The same information may reveal the potential for a new risk that has not occurred but can be planned for. It is important that the project manager and those who help analyze work activity information understand that information not only documents what has happened, but also can inform those who are looking that a possible risk event may be imminent. This is a very important element of activity monitoring because it gives the project manager and project staff the rare opportunity to plan for a potential problem before it happens.

New risk potential typically is discovered in analyzing project activity data. This is the point at which the project manager or project staff need to be knowledgeable and pay attention to what the information is actually telling them to detect a problem that has not already occurred. This may be an opportunity for the project manager to train other project staff who assist in analyzing information from monitoring work activities. He can help them see how the information can show the difference between problems that have already occurred versus a problem that has the potential to occur in the future. This type of training is invaluable for project staff because it maximizes the use of information gathered from work activities.

Corrective Action Requirements

Based on the outcome of information that has been gathered and analyzed from project work activities, determinations can be made as to whether a work activity is on budget and on schedule and requires no further action. Determinations also can be made whether some form of problem is influencing the activity that needs to be corrected to keep a work activity on schedule and on budget and within the quality expected. When the project manager prepares and communicates work performance reports that indicate a problem is occurring on a specific activity, these reports typically generate discussion and further root cause analysis to determine what can be done to correct the problem. The outcome of these discussions and analysis will, one hopes, produce a *corrective action requirement* that is documented and implemented on the work activity. This is another important reason that work activities are monitored and information is gathered and analyzed. The purpose is not only to understand the status of work activity, but to provide information the project manager can use to "manage" the quality of work performed and to ensure activities are on schedule and on budget. More about corrective action control of work activities is covered in Chapter 12, "Schedule and Cost Control."

Forecasting Adjustment Requirements

As the project manager monitors work activity and reviews and analyzes information gathered on activity progress, either work activities are on schedule and on budget, producing the expected quality of work, and require no further action; or the information suggests a problem or trend that requires corrective action. Such action can impact either the budget or the schedule, so adjustments in forecasting the remaining work activities in the project plan may be required. Forecasting is based on the original project plan, which includes estimates of schedule duration, estimates of work activity costs, and definition of work to produce the expected deliverable, which formulates the baseline of the project. This is the expected plan the project manager follows and uses as a baseline to compare actual work, activity duration, and cost to measure activity progress.

If corrective action or controls that have to be implemented have an unavoidable influence on the schedule or budget, the project manager has to make adjustments and publish updated expectations for completion. If adjustments influence the schedule, the updated forecast is derived from the variation of actual duration compared to the baseline called schedule variance (SV), which can be calculated and communicated in the estimate to completion (ETC). If adjustments influence the budget, the updated forecast is derived from the variation of actual costs compared to the baseline called cost variance (CV), which can be calculated and communicated in the estimate to completion (ETC). The project manager then needs to effectively and efficiently communicate the updated forecast or estimate to completion to the project staff and other individuals within the organization who need to know critical updates to the project plan.

Change Validation Analysis

We have discussed the importance of monitoring work activities to collect information on the quality of work, duration of work activity, and cost of work activity to compare against a baseline as to the progress and status of work activities. Monitoring also plays another very important role: in the event a change to a work activity is implemented, the effects of that change also need to be monitored to validate the change is producing the desired outcome. Just because a change is implemented doesn't mean it always works as anticipated, so monitoring the activity with the change implemented produces new information that can be analyzed regarding the effect the change has produced.

The reason for monitoring work activity is for the continued collection of information for activity progress, and this is an ongoing process until the work activity has been completed. Changes to work activities can produce other anomalies or new problems, or they can introduce a potential risk. Monitoring the effects of a change on a work activity is very important because the change can produce either the positive effect that was anticipated or a negative effect that creates more problems. If a negative effect is produced, the project manager must know immediately so that information can be analyzed and a new course of action can be taken quickly to mitigate or eliminate further impact to the schedule or budget.

Monitoring project activities is the means by which a project manager derives information and analyzes what is actually happening in comparison to what is expected as documented in the project plan. Through the implementation of monitoring tools and techniques, the project manager can "manage" activities instead of simply reporting on status.

Review Questions

1. Discuss what is meant by integrated monitoring.
2. Explain the primary reason that project managers monitor project work activities.
3. Discuss how information-gathering tools would be implemented on a project. Use your own sample project.
4. Why is information analyzed, and what specific pieces of data would be of interest to the project manager?
5. If analyzing activity information reveals a problem, explain why a root cause analysis is needed.

Applications Exercise

JP Phentar Construction: Case Study

JP Phentar has owned and operated his construction company for 27 years and is currently interested in building a custom home for his own family. Phentar Construction has specialized in large and exotic custom homes built in areas that present challenges, such as heavily wooded and rocky terrain, hillsides, and sandy beach sites. Phentar Construction has built large custom homes for executives, heads of state, and movie stars around the world; those projects generally include interesting and challenging amenities for construction companies to manage in the course of building homes.

Phentar is pulling out all the stops on this construction project to include things in his own home that his family can enjoy although they are typically out of the norm for most family residences. Phentar has purchased three acres of foothill terrain that include several large rock outcroppings that have to be removed for the construction of an 8,500 square foot six-bedroom, six-bathroom home. This home will also include an elaborate game room with professional pool table and arcade games, a large family room with rock fireplace, and a fully functional home theater room with large-screen TV and surround sound and theater seating. The home will also feature as its primary centerpiece a 35,000-gallon saltwater fish tank that will start at the main floor in the center of the house and extend for two stories to the ceiling. This fish tank will include a large rock wall covered with all manner of coral and sea urchins, flowing water movement, and it will be stocked with an elaborate display of tropical fish. The exterior of the home will include a large pool with spa and a covered patio with full outdoor kitchen, including a fully functional brick fire oven.

Because Phentar has built homes with similar amenities in the past, he knows all too well several of the contractors required to outfit these types of amenities can present challenges in cost estimation, level of quality, and ability to stay on schedule. Concerns with this particular project lie within clearing the initial acreage of large rock, and specialized amenities such as the game room, theater room, and fish tank that can present challenges during the scheduling of these activities in the course of building the home. There can be serious issues in the timing of these activities because they can affect other areas of the home during construction. Phentar is confident this project can be completed if proper project management tools and techniques are implemented to monitor and control critical activities through the course of this project life cycle.

Case Study Exercise for Chapter 11

1. Determine what information-gathering tools would be most effective on this project and what activities would need to be monitored.
2. Based on data that would be generated from work activities, what types of analytical tools could be used to determine project status?
3. Can any corrective actions be initiated?

12

Schedule and Cost Control

Introduction

When projects are approved within an organization, the expectation is that the resources and materials procured per the original estimate are sufficient to complete the project objective as outlined in the project plan. Depending on the size and structure of the organization and the size and complexity of a given project, the project manager may be responsible for or assist in the development of the project plan, schedule, and budget. However, the project manager's *primary roles* are *monitoring project activity* and *implementing controls* to ensure project activities produce the expected deliverable on schedule and on budget. Gathering and analyzing work activity information and assessing the need for the implementation of controls consume a great deal of the project manager's time.

It is important to understand the difference between *monitoring* and *controlling* project activities. Monitoring, as discussed in Chapter 11, "Schedule and Cost Monitoring," is information gathering and analysis to determine status of work activity performance. It is strictly an information-gathering and analysis function to determine whether actions need to be taken to produce corrections in work activity performance. It's a very important role the project manager plays, because this is the first step in determining whether projects stay on schedule and on budget and produce the expected quality in the project deliverable. The second most important task the manager conducts, based on the analysis of information gathered, is to determine whether changes or controls need to be implemented to make corrections in activity performance. So monitoring and controlling project activities can be summed up in the two primary functions:

Monitoring—Information gathering and analysis to understand *what is happening* on work activities

Controlling—The *action taken*, based on the analysis of information gathered, as to corrections (changes) that will be implemented to adjust work activity to yield the expected activity performance

The characteristics of control are much different than monitoring in that control is an action taken to implement a change to influence a situation, whereas monitoring does have action associated with it; these actions are limited to gathering and analyzing data for the determination of work activity performance status. The results of monitoring determine whether an activity is on schedule and on budget, or require actions to be taken that will impose change to influence work activity performance. If the intent in managing project activities is to ensure the quality of project deliverables and that schedule and budget are maintained, the project manager has to control work activities to ensure performance is maintained, and this may require her to implement changes.

Change Control

The primary component of maintaining control of work activities is implementing changes to influence activity performance. If change is the primary tool used to maintain control of work activities, the focus should be on controlling change and understanding the impact change can have on a project. Change, for the project manager, can be both good and bad because the intent of what a particular change is trying to accomplish can be good, but if the implementation of change is not controlled, the outcome of that change can be unpredictable.

Changes can be made at several levels within a project. For example, changes may involve scope and expectations of the project deliverable, organizational processes conducted on work activities, stakeholder expectations, and any number of issues that may have to be addressed with resources allocated to the project. It is common for changes to be made with regard to procurements and the contracting of outside resources required for work activities. In some cases, critical changes might have to be made to documentation to define specific processes for work activities or sent to vendors or suppliers who will characterize items for procurement. If a change on a project is required, the project manager should develop and utilize a change control process, if one is not already developed within the organization, to manage its implementation.

Change Control as a Process

Like so many other areas of work activity throughout an organization, change control activities are conducted by means of developing and maintaining processes that define each step required to perform an activity correctly and efficiently. The way to conduct changes on a project is simply an outline of activities required to implement those changes and therefore can be managed as a process. The only way change can truly be controlled is by developing a change process so that it can be documented and utilized with consistency. The project manager can then use this process throughout the project at whatever level required in managing the implementation and measuring the effect that change will have on project work activities.

The four primary steps that make up the process of change include *propose, implement, communicate,* and *measure.*

1. **Propose**
   a. **Gather data.** The first step in determining whether change is required is to review information gathered from work activities, analyze it for compliance to work activity expectations, and decide whether corrective action is required. Changes should always be based on actual data taken from work activities and not from opinions or hearsay of individuals who are not involved directly with the work activity in question. It is important the analysis of information clearly indicates the need for correction and that there is no question as to the validity of what the data indicates.
   b. **Define the need for a specific change.** After it has been determined that correction is needed for a work activity, that information and analysis can be used to develop the details and scope of a proposed change. This step in the change process is important because it summarizes what the data actually indicates in a form that can be understood by those who will be evaluating the need for the change. It is also important to include what the expected outcome will be on activity performance.
   c. **Propose change.** After a need for change has been clearly identified, the change needs to be articulated and presented in the form of a proposal. This can be in the form of a short statement outlining details of the change, called a *change order form.* Or it can be complex and detailed, requiring a full written proposal outlining all the details, charts, graphs, and other supporting documentation to accurately articulate the scope of change. The proposal's development is the final form indicating how the details of change will be documented and communicated to those evaluating the change for approval.

d. **Validate and sign off.** The proposal should be submitted to a team of individuals who are knowledgeable in the work activity such that they can evaluate the information provided in the proposal and determine whether this is an acceptable course of action. It may be determined that offline testing should be required to validate the outcome of a proposed change. In other cases, the opinions of subject matter experts regarding other courses of action may produce a similar outcome with less risk or impact to schedule or cost. After the change has been validated, it is important that the team sign off on this document to validate that a change has been evaluated and approved; this information can then be communicated in project status updates.

2. **Implement**
   a. **Conduct changes.** After a proposed change has been approved, implementation can be a difficult hurdle to get over, depending on the type of change required. In some cases, there may be resistance because it is common for people to reject change. The reason can be simple fear of the unknown or a lack of details regarding how the proposed change will produce the expected outcome. It is best for the project manager to inform in advance those staff who will be implementing the change so that they can ask questions, and details can be conveyed to help them understand how and why a change is being implemented. It is also important for the project manager to convey that work activity performance is noncompliant and that the change is necessary to bring the work activity performance back in compliance to the baseline expectations. The project manager should understand the importance of gaining the trust and the buy-in of resources who are implementing the change because this can be an important factor in the success of what the change is trying to accomplish.
   b. **Manage scope of change.** The project manager and/or the responsible individuals tasked with implementing the change need to manage the details that define the scope of the change intended. This task is important because those implementing the change may interpret some of the details differently than those who understand the scope of the change more clearly. The change will be successful only if implemented exactly the way it was documented and proposed, so careful management of details during the implementation is critical.
   c. **Publicize changes.** When the implementation process is complete, the project manager and/or responsible individuals implementing the change should document that all steps have been completed and the process of this change has been verified. This task is important so that staff conducting the work

activity and other project staff and stakeholders know when the change has been accomplished. It is important that the change process have a definite completion point so that information gathering and analysis can be documented from that completion point and reflect the impact that the change was designed to make. It may also be a requirement of the change process to have a final signoff officially validating that the change has been completed.

3. **Communicate**
   a. **Establish who needs to know.** As with the initial group of individuals required to evaluate and sign off the proposal, another group of individuals who have interest in knowing that the change is complete needs to be established. This group may include the original individuals evaluating the proposal but might also include functional managers and executives or other supporting staff who need to know the change has been implemented.
   b. **Determine appropriate method of communication.** Depending on the type of information used in the original proposal and the information gathered to validate that a change has been completed, individuals receiving this information need to have it in a form that can be easily understood. If the change can be easily articulated in a memo or email, this method can be used for simple forms of communicating work activity status reflecting a change. If a change is more complex and requires a much more detailed and sophisticated proposal, similar levels of detailed information need to be prepared in forms that can be effectively communicated to other individuals. The communication of information may not be within the same location, so other creative forms of communication might have to be used if individuals are in other locations.

4. **Measure**
   a. **Compare to original baseline.** Another important component in the implementation and completion of a change for a work activity is to continue measuring the activity and analyze performance as compared to the original baseline. This step is required to validate a change is actually producing the expected outcome and work performance is being measured in compliance with the project baseline. With any change, it is required to validate the success of that change, or to determine whether other problems were created or the change simply did not produce the expected outcome and should be reversed. It is important to note that simply making the change does not always improve work activity performance and, in some cases, may actually create other problems. It is extremely important the project manager

validate that a change produces the expected outcome because this is the control function she uses to bring work activity performance back into alignment with the project baseline.

    b. **Determine sustainability.** The last component of the change process needs to be a determination of the sustainability of a change. In some cases, changes have a permanent influence on work activity performance, and the change could be considered permanent and sustainable. In other cases, changes might have a temporary influence, but the ongoing measurement and analysis of data show fluctuations in performance and question the sustainability of the change. This data is very important because the project manager needs to analyze and understand the validity of a change and whether or not that change should be kept in place or reversed. In some cases, a change might simply need some minor alteration to improve its outcome and sustainability. This again points back to the importance of monitoring, gathering, and analyzing data on work activity performance to track any fluctuations or variance in performance after a change has been implemented.

Integrated Change Control

The change control process is a generic outline that can be used at any level throughout the organization, but other project elements have to be considered for a change control process to work effectively. *Integrated change control* takes the change control process and integrates it within the project to utilize the original statement of work, established project baseline, identified project risks, and predetermined contingency plans already established within the project plan. By designing a change control process and integrating it within the project, the project manager achieves structure and control over the change control process, but also customizes that process to specific attributes of the project to gain maximum control capability.

Other documents and artifacts that can be used for integrating change control into a project include the project management plan, change requests and proposals, and work performance information and analysis. The other unique element of integrating the change control process within the project is the ability to solicit expert information from project staff directly associated with work activities that can expedite information gathering and analysis for change proposals. Another advantage of change control integration is the ability to utilize expert information from project staff to assist in change verification because this is the most important element of controlling change. Having this process integrated within the project allows the project manager even more control over this aspect of change.

Integrating change control within a project also allows the project manager a more expedient communication of status in the course of a change; it is easier to communicate immediate status of change on a work activity with those more closely related to that activity, thus giving the project manager another element of control in the change process. The project manager also uses the change control process in the event that adjustments to the baseline are appropriate and need to be properly documented. The overall goal in integrating the change control process is the efficient and effective identification of a change requirement, development of a proposal and approval process, and implementation and verification of changes.

Control Tools and Techniques

Much like the monitoring tools that gather and analyze information, control tools and techniques focus primarily on the actions that will be taken to redirect project activity to improve performance that aligns with the project baseline. It is important to understand that control tools and techniques do not always incorporate analysis functions, but they suggest actual actions the project manager can take to improve control.

In the following sections, we cover several tools and techniques that can be used in the control of project work activities. Some techniques can be used to control both schedule and cost components of work activities, whereas other tools and techniques are used more specifically for schedule, cost, and quality control. Some tools covered here were also covered when developing the schedule and budget because they can be used in both development and control. In the selection and implementation of control tools, it is assumed that as a result of monitoring work activities, information gathering and analysis indicate that change is required and a control tool needs to be utilized.

Because the project is composed of three primary areas—schedule/time, cost, and the quality of the deliverable—these three items make up the *triple constraint*, described previously. As the project manager initiates project activities and monitors activity to derive information to analyze work activity performance, the ultimate goal is to maintain work activity to a baseline schedule, budget, and expected quality for the project deliverable. The project manager needs tools to conduct control actions that influence project activities to maintain a baseline schedule budget and level of quality.

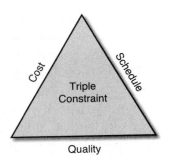

Figure 12.1 Triple constraint

Contingency Control

In developing the project management plan, the project manager typically considers contingencies in both schedule and budget to create either buffers or padding, or designed for risk event planning. In most cases, contingency planning is typically performed at the beginning of a project. This way, if contingencies are required, they can be allocated as part of the baseline. On the rare occasion that a new risk event might be identified in the course of a project, a new contingency can be applied and even included in a baseline adjustment through the change control process.

The biggest control element in having contingencies planned throughout the project life cycle for either schedule or budget concerns is in managing how and when they are used. The reason to use a contingency is control, so one should be used only in the event that control is absolutely required. If time buffers are designed within certain paths of a network diagram, they should be used only if required to stay on schedule. It is critical the project manager use discretion in the communication of schedule padding for work activities, because those performing the work will use the extra time if they are aware it is available. This is how the project manager controls contingencies by communicating their availability only when necessary and controlling their use. Schedule contingencies can be a valuable tool if used at the right time for the right purpose, but it is incumbent on the project manager to perform root cause analysis if information suggests a work activity is over budget or falling behind schedule. The warning is not to simply throw contingency scheduled buffering every time something falls behind schedule. The way to control contingencies is to use them only how they were designed and only if doing so is the last resort.

Schedule Control

The first element of the triple constraint the project manager needs to control is the actual time spent conducting a work activity. Depending on the size and complexity of a project, this task can be very simple given a simple work activity, or it could also be incredibly difficult given an extremely complex work activity. The project manager must remember that control is required only if the information gathered on a work activity has been analyzed and suggests that a parameter of activity duration, cost, or quality has shifted beyond acceptable limits. It is also recommended that a proposal be established, reviewed, and approved regarding the form of control that would be implemented because a control typically requires a change of some kind. Using the tools and techniques to control schedule described in the following sections, the project manager can improve her schedule in several ways, depending on what's available within the organization and within project limitations.

Data for Schedule Control

The first place the project manager looks for information to control schedule is in the monitoring process. This is the place where information is gathered and analyzed, concluding that a problem within a work activity has increased activity duration. This analysis also uncovers the root cause of the problem, providing data for the project manager to solicit possible solutions to control the work activity duration. It is recommended that the project manager use only the data gathered directly from the work activity and analyzed for the basis of considering a change to control work activity duration. The reason is that this is the first-hand information available that would suggest a problem needs to be addressed. For schedule-related problems, it is typically best to focus attention on parameters and characteristics within the specific work activity because these are typically where root cause analysis reveals problems. The project manager can then solicit advice from subject matter experts and others within the organization regarding possible solutions, but the analysis of how to fix the problem should always be based on first-hand information gathered directly from the activity creating the problem.

Critical Chain Method

The *critical chain method* (CCM) is a control tool that implements buffers within a network of project activities at specific locations to control the project schedule. The critical chain utilizes the critical path approach of network diagramming project activities and is most effective in resource utilization and optimization. Buffers are represented similarly to activity items connected within a path that are identified as nonwork activities but have duration. The primary philosophy in the use of buffers is to balance pathways to accomplish corrections in scheduled convergence to other pathways. Buffers can be utilized in two primary areas within the network of activities:

> **Feeding buffers**—These buffers are placed on specific paths that need influence to control the cumulative schedule durations for a particular path.
>
> **Project buffers**—These buffers are placed on the main or critical path to control the overall duration of the project schedule.

An example of the use of both feeding buffers and project buffers is shown in Figure 12.2. Here, the convergence of two pathways into the critical path might have to be adjusted using buffers. A buffer might be placed in the main critical path toward the end of the chain of activities to help protect the finish date and stay in compliance with the baseline of duration estimates and the expected project completion date.

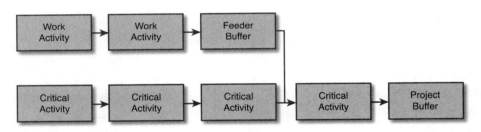

Figure 12.2 Critical chain method

Resource Leveling

One of the most common issues measured on work activities is the utilization of resources and how certain resources can present challenges or constraints. As a reminder, resources can be human resources, capital equipment, facilities, available finances, and contracted equipment or services brought in from outside for use on project activities. Because all these types of resources are identified and scheduled at the beginning of the project, their availability and effective utilization at the time they are required on a work activity may not always go as planned.

In most cases, overutilization and changes in availability of resources are typically the root causes of problems that create either schedule or cost-related issues when work activities need to start. *Resource leveling* is a technique typically used during the initial scheduling process whereby the project manager can make adjustments in resource utilization not based on one work activity alone, but taking into consideration several work activities within a particular duration of the project. The project manager can make adjustments in the utilization of resources to level out the number of hours worked per day and the number of resources required per day per work activity.

Resource leveling is most effective in resource-constrained situations. For example, critical resources are available only at certain times or in certain quantities; resources have changed to being overallocated in demand loading; or it is determined that previously scheduled human resources may not have the required skill set, so either additional resources need to be added or skilled resources need to be acquired from other projects. These problems can be typical and common for project managers on all types of projects—large or small, simple or complex.

Resource leveling can also be used as a *control tool* on work activities where resources have changed and need to be releveled to solve a resource overallocation problem. As discussed in Chapter 8, "Schedule Development," leveling can be done in two ways: adding more resources if a time constraint exists or adding more duration if a resource constraint exists. Figures 12.3 and 12.4 show how leveling a resource-constrained activity would look when more duration is added.

Resources	Electrical Engineering Resource Availability Change																
	1	2	3	4	5	6	7	8	9	10	11	12	13	14	15	16	17
A. Develop Requirements	RS EE	RS EE	RS EE														
B. Design Housing Components				ME	ME	ME	ME	ME									
C. Design Sub-Assembly A				EE	EE	EE	EE										
D. Design Sub-Assembly B				EE	EE	EE	EE	EE	EE	EE	EE						
E. Design Sub-Assembly C				EE	EE	EE	EE	EE	EE	EE							
F. Assemble & Test B & C												AS EE	TT				
G. Final Assembly														AS EE	AS		
H. Final Test																TT EE	TT
Available Resources: RS =1 EE = 2 ME = 1 TT = 1 AS = 1	RS 8 EE 8	RS 8 EE 8	RS 8 EE 8	ME 8 EE 24	ME 8 EE 24	ME 8 EE 24	ME 8 EE 24	ME 8 EE 24	ME 8 EE 16	ME 8 EE 16	EE 16	AS 8 EE 8	TT 8	AS 8 EE 8	AS 8	TT 8 EE 8	TT 8

Figure 12.3 Resource-constrained activity

Resources	Resource Requirement after Leveling																		
	1	2	3	4	5	6	7	8	9	10	11	12	13	14	15	16	17	18	19
A. Develop Requirements	RS EE	RS EE	RS EE																
B. Design Housing Components				ME	ME	ME	ME	ME											
C. Design Sub-Assembly A				EE	EE														
D. Design Sub-Assembly B						EE	EE	EE	EE	EE	EE	EE	EE						
E. Design Sub-Assembly C							EE	EE	EE	EE	EE	EE	EE						
F. Assemble & Test B & C														AS EE	TT				
G. Final Assembly																AS EE	AS		
H. Final Test																		TT EE	TT
Available Resources: RS =1 EE = 2 ME = 1 TT = 1 AS = 1	RS 8 EE 8	RS 8 EE 8	RS 8 EE 8	ME 8 EE 16	ME 8 EE 16	ME 8 EE 16	ME 8 EE 16	ME 8 EE 16	EE 16	EE 16	EE 16	EE 16	EE 16	AS 8 EE 8	TT 8	AS 8 EE 8	AS 8	TT 8 EE 8	TT 8

Figure 12.4 Resource leveling, adding more duration

Schedule Crashing

In the event that resource leveling or smoothing may not be an option, another form of schedule control is to affect the actual duration of a work activity with the use of added resources; this is called *schedule crashing*. With this technique, the project manager simply increases the amount of a particular resource on a work activity to improve the duration of that activity, and this typically requires added cost to the project. The key feature of this technique is selecting the least costly resource that can provide the greatest amount of reduction in schedule duration. Specific examples of crashing described in Chapter 8 show how to utilize budget to reduce duration if activities are going over budget and seem to be getting out of control.

This method is especially effective as a control technique because activity duration can be greatly reduced if financial resources are available to add resources to complete work activity tasks if the activity is time-constrained. Crashing should be used only in controlling project schedule if no other options are available because it usually adds to the budget.

Fast Tracking

If it is determined that the duration of a sequence of activities is too long and needs to be reduced to maintain schedule, another form of schedule reduction within a particular path of work activities is called *fast tracking*. This technique is used to compress the overall duration of a network path by shifting work activities from being in serial (see Figure 12.5) to performing work activities in parallel. The project manager needs to review the predecessor/successor relationships and dependencies of neighboring work activities to determine whether performing two activities in parallel would be an option. This is an excellent way to reduce the schedule of a particular path at no added cost to the budget. An example of fast tracking is shown in Figure 12.6.

Figure 12.5 Standard serial network

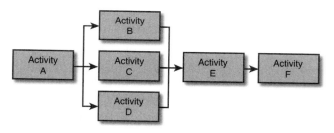

Figure 12.6 Fast tracking a network

Cost Control

The second element of the triple constraint has to do with controlling the cost of work activities. This task can be much more difficult for the project manager because several individuals or departments may be involved with cost elements of project activities. The procurement of materials and possibly the contracting of external resources can also play a role in influencing the actual costs incurred on work activities. Because the project budget was developed primarily from estimates that have been gathered, contracts that have been signed, or information based on historical cost, these details are all gathered and documented at the beginning of the project. Depending on how long the project life cycle spans, they can play a role in the accuracy of these estimates over time. The only true estimates are those signed in contracts or based on official quotes from vendors or suppliers that, if used within the time frame of the quote, should be valid. These cost control items typically can be engineered at the beginning of the project to ensure these costs do not change when they are incurred within the project.

Project managers must also be aware that individuals within the procurements department may introduce organizational processes that were not accounted for when original estimates were made, and these processes may influence the actual price paid for certain items that have to be procured. Other areas of control may include individuals tasked with the purchase of items, so clear expectations have to be communicated regarding what the actual costs need to be limited to in order for work activities to stay on budget. If man-hours are being tracked and the project factors this information into the baseline of work activity costs, the project manager needs to oversee or "control" the amount of work being performed so as not to go over budget; this also can present more problems in the type of labor and what is actually being accomplished in the work activity.

The project manager must be aware of all the things that can influence cost on work activities. She must monitor them as they are happening to ensure problems are not occurring and controls do not have to be implemented. In some cases, the project manager can review procurements and expenditures prior to their happening in a form of control to ensure actual costs match or are less than the budgeted estimates for those items. It is the project manager's responsibility to oversee all costs incurred on a project and to review the information gathered for procurements and expenditures on work activities to assess whether controls are necessary.

Data for Cost Control

Initially, the project manager or project staff should review information gathered and analyzed during the monitoring of work activities to determine whether cost controls are required. If fixed-cost items are procured based on quotes from suppliers or vendors, and contracts have been signed establishing a fixed price for a resource, material, or service, then these items do not necessarily need much control relative to change, but more so the oversight of execution to ensure terms of the agreements have been fulfilled. In other cases where the procurement of materials or resources has been identified and only estimates have been derived, the actual price may fluctuate; this requires controls to ensure procurements stay within the budget. This information can be found primarily in the work activity requirements of cost estimates because this is the place where the original estimates were derived and the agreed-upon fixed pricing or simple estimates were recorded.

Other areas the project manager or staff can derive information for cost control might be in the accounting department, which will no doubt track all expenditures recorded for each project within the organization. If there are questions as to whether procurements are within budget, this would be the department to clarify that type of information. If certain work activities or areas within the project are over budget, the project manager can find out more precise detail from the accounting department as to what will require controls. The procurements department is helpful in informing the project manager about what procurements might have cost fluctuations, and the project manager can introduce limitations as a form of control to guide procurements in staying within budget.

The project manager also can use historical information to locate items that have been purchased on other projects, and it may assist her in understanding whether controls need to be implemented based on prior purchases. Soliciting information from subject matter experts and others familiar with a particular work activity may also reveal details about specific items that need to be procured to understand whether controls are necessary.

Procurements

The procurements department is established to oversee and conduct all purchases required by the organization. It more than likely has established several processes regarding the way information is gathered, documented, and utilized to perform procurements. Although this department is typically not a part of a project, it performs many functions within the project life cycle to acquire resources and materials to accomplish the project objective. The project manager must understand the importance of procurements control because this is one of the primary areas she controls in the budget. As previously indicated, some items identified for purchase are based on a fixed price listed within a quote or a contract that, when initialized, should produce the same result as documented within the baseline budget. These are the best-case scenario procurements because they have already been committed to suppliers or vendors and, upon purchase, will result in on-budget performance.

Other procurements identified within work activity requirements may be estimates derived by any number of means with varying degrees of reliability and accuracy. The project manager needs to pay close attention to controlling these procurements because they are the most likely to have variances from the budget baseline. The project manager must work with the procurements department before these purchases are made to ensure what is being purchased matches the requirements within the work activity. If the project manager can understand the characteristics and functionality before the procurement actually takes place, she can implement controls. The key element here is monitoring when procurement requirements are scheduled and that the project manager acts before they are made to verify what will be purchased.

The project manager must also be aware that original estimates gathered for a work activity simply include the costs of required items. These costs might not reflect taxes, shipping, expedite fees, or any other fees incurred in the procurement of each item. This is where the project manager can confirm that the estimated cost of an item is accurate to what the procurements department is proposing to execute but does not include any of the preceding items that will drive this cost to a variance over budget. (This detail is covered in Chapter 9, "Cost Estimating.") Another form of cost control that can be implemented on a project is ensuring that all costs required to procure an item are included in the estimated value so that when it is time to procure that item, there are no surprises as to the total cost that will be realized.

Contracts

Project managers also need to pay attention to controlling costs when resources require contracts to execute their procurement. Contracts are legally binding agreements between a party supplying a product or service and a party providing compensation after receiving the product or service. Although this concept sounds fairly simple, several types of contracts that can be drawn can have any number of effects on the overall cost of a product or service, and in turn, this can have an impact on a project and an organization. The most important element in negotiating contracts is the balance of benefit versus risk for each side and what each side is willing to offer or give up in adjusting that balance to establish an agreement.

Because there are many types of contracts that are explained in great detail in other texts specializing in contracts for projects, we focus on three different contracts typical in project management that have varying degrees of risk. These contracts are not listed in the way they are structured or used, but more in the way project managers might expect to use controls to ensure costs stay on budget when using a contract:

Fixed-price contracts—In these contracts, a documented product or service is produced and delivered at an agreed-upon lump-sum fixed price. In this scenario, suppliers assume more of the risk because they have committed to the delivery of a product or service, regardless of their cost to produce that deliverable. Receivers of this product or service operate at a lower risk because they receive the deliverable at the agreed-upon price with no risk of a price increase. Little control is required in this scenario because this agreed-upon price should be listed in the original work activity requirements and documented within the project baseline. After the contract is signed, the agreed-upon fixed price should meet the expectation within the project baseline.

Fixed-price incentive fee contracts—These contracts work similarly to fixed-price contracts because there is an agreed-upon product or service and a fixed price for the initial product or service. But the addition of an incentive clause can shift the balance of risk back to the supplier because this allows for increased profits to meet the conditions of an incentive. For example, an incentive might be a bonus if the supplier delivers the product or service earlier than scheduled. The supplier still has to produce the product or service, but there is an increased profit if it is delivered early, which benefits both the supplier and the receiver of the deliverable. Because both the fixed price and incentive can be well documented in the original work activity estimates, the project manager can include the incentive within the project baseline budget as a buffer or contingency for early completion, and this accounts for not requiring controls.

Time and materials contracts—These contracts work to the benefit of suppliers because they are simply paid for time expended to produce a product or service and the invoicing of materials required in the agreed-upon deliverable. This shifts the balance of risk to the receivers or customers because the suppliers have no incentive to complete a deliverable in a faster period of time versus a longer period of time and are simply reimbursed for the cost of the materials. This type of contract requires the greatest amount of monitoring because contractors may have a tendency to take longer to complete a work activity deliverable to increase the revenue gathered for this particular contract. The project manager must understand that controls might be required to limit the amount of time and materials used to accomplish an agreed-upon deliverable so as not to go over schedule or budget.

Quality Control

The third primary area of control is in the creation of the project deliverable itself. Each work activity produces a documented deliverable to accomplish the expectation of that work activity. Documentation shows what materials will be used, the amount of time that should be taken to produce the deliverable, and the expected characteristics and functionality designed in the requirements of that work activity deliverable. The measurement of how a deliverable is produced can be defined by its *quality*. Quality is meeting the expectation of the customer for a product deliverable and is typically defined in two general terms:

Intended functionality—This term refers to the production of a deliverable that accomplishes all the originally designed characteristics and functions required by the customer. When a deliverable is evaluated against original specifications, drawings, or documented intentions of functionality, the quality standard is that it meets or exceeds the expectations of the customer receiving the deliverable. This characteristic of quality should be built in at the beginning of the project when discussions and negotiations between the organization and the customer define the project deliverable. The more detail in this process, the better the ability to understand and control work activities to produce a deliverable that will meet customer expectations.

Materials and workmanship standards—This term refers to the types of materials and the level of expertise of human resources, machines, or equipment used in producing the deliverable. This criterion is different from functionality because a deliverable can meet all the functionality expectations but be created out of materials that do not meet the customer expectations. This can also be referred to as a deliverable meeting all the functionality and created with materials that meet customer expectations, but created by human or equipment resources that do not produce an end product that meets customer expectations. This measure of quality has to be defined in the work activity requirements regarding specific materials and the form of workmanship standard that is acceptable by the customer. The project manager needs to monitor these criteria to gather and analyze information to determine whether a work activity meets customer expectations.

Quality Inspections

Depending on the type and structure of an organization as well as the typical deliverables produced and projects within that organization, various inspections can be designed to qualify or quantify the production of work activity deliverables. If a deliverable is being produced within an organization, quality inspections can be implemented as a control feature to stop further progress on a product deliverable if it is determined that the progress is not meeting quality expectations. These can be called *quality inspection gates* or even considered *milestones*, depending on the size and complexity of the work activity deliverable.

It is also important that the project manager uses some form of quality measurement to determine whether work activity has the potential to meet customer expectations. This can provide a *monitoring function*, in which information gathering and analysis are performed during the quality inspection, and a *control function*, in which the progress of a work activity can be halted if it is determined that work is not meeting quality expectations. This is an extremely valuable control tool for the project manager because it can be assigned to an individual who is knowledgeable and skilled with a work activity and has a clear understanding of customer expectations and therefore can perform a quality inspection and initiate controls if required.

Regulatory Inspections

Regulatory inspections are similar to quality inspections but are initialized through requirements of a local, state, or federal government agency in conjunction with a permit that was issued or requirements based on specific characteristics of the deliverable. An example is the construction of a building. Plans that are drawn outline all specific characteristics regarding how the building will be constructed, what materials will be used, and what specific processes are required. Upon evaluation and approval of these plans by a regulatory government agency, a permit will be issued to commence construction. In this permit, stop points or inspections will be required before further progress can continue; they ensure completed components of the work activity meet the expectations outlined within the approved permitted documents. After an inspection is completed and work is approved, the next phase of work activity can commence. These inspections give the project organized and regulated controls designed by external agencies to oversee the control of work activities.

Design Reviews

Projects within an organization that utilize engineering resources or development-type resources can use a process called *design reviews* and stop points during the production of items like prototypes. These types of project deliverables are broken up into major phases and have milestones or stop points where engineering, subject matter experts, and management staff can evaluate the progress of work activity. Design reviews are another form of control in which information on work activity performance is gathered and analyzed, but work activity is halted upon review of the analysis to determine whether alterations have to be made or whether work can proceed as planned. The fact that work activity is halted for analysis is the actual control component of the design review.

The project manager can have design reviews built into the work breakdown structure to stop work progress for periodic review. She also can include the customer in the design review process to determine whether the next work activities need to be altered or work can commence as designed in the project plan. A design review is an excellent form of control in that an entire project deliverable is not created; this way, at the end of the project, it will not be determined that the customer's needs have changed and significant alterations to the deliverable have to be made. Design reviews allow the customer to evaluate smaller steps in the development of the project deliverable to make small alterations through the change control process; these steps can be well documented and even included in the project baseline. Design reviews present an excellent form of control in not only controlling work activity but also taking into consideration customers' needs and changing requirements.

Control Results

Correctly implementing control should result in monitoring and measuring work activity performance that has been influenced or changed by the change control process and is now performing as expected according to the schedule, cost, or quality baseline. Although the final result is simply the correction of a characteristic in work performance, as you have seen with areas of monitoring in Chapter 11 and with tools and techniques that can be used to implement change to control a work activity, there is a great deal more involved in correctly controlling performance. It is important the project manager and the project staff understand the steps involved; the tools and techniques in monitoring, gathering, and analyzing information; and the conclusion of root cause analysis; as well as control techniques that can implement a change control process to modify performance. If the monitoring and control processes are implemented correctly, the project manager has real control over a project and can manage work activities to a project baseline.

Reporting Controls

Part of managing work activities involves reporting on the status of work activities and, as you have seen in this chapter, reporting any change or the results of change on work activity performance. Because the project manager, through the monitoring process, gathers and analyzes information to draw conclusions on work performance and actions required, so other management and executives require information from the project manager on the status of the project and work activities. In some cases, they may need this information for accountability to ensure that the project is on schedule and on budget, but in other cases, reports may simply describe work activity performance.

Although the project manager typically prepares reports for functional managers, a program director, or other management executives, it is important the project manager understand what type of information needs to be reported, to whom, and what level of detail is required. Executives typically do not want to see much detail, but they want truthful and accurate accounts of project status so that they can assess the overall business plan and objectives of the organization. Other functional managers, project staff, quality and engineering departments, and accounting staff may want more detailed assessments of specific work activity because they are interested in how the project is progressing and exactly where activities are within the budget and the schedule. Reporting to this level would also be in the best interest of the team, to discuss current problems, issues, and contingency plans of potential risks. The project manager must again be aware of the audience to whom the reports are going, what information they need, and what level of detail is required.

The project manager should always make it a practice to document everything that is happening on the project including problems and issues on work activities, outcomes of change control processes, and the overall strategy she has taken in completing a project deliverable. This valuable information can be archived and used on future projects, for the development of project management plans, and historical data for lessons learned and/or ways to run a successful project. In some organizations that have a project management office (PMO), this type of data and information gathered from projects should be archived and organized in a standardized form that can be used by other project managers.

Manage Change Control

As you have learned in this chapter, it is important the project manager understand how the change control process works in "controlling" project work activity performance. Because the change control process itself represents an organized way to manage the process of change within a project, the project manager should use this process to effectively and efficiently manage control. Change is a welcome and good thing if it is designed well, implemented, measured, and validated in an organized and controlled process. One of the primary components of controlling a process is the occasional need for change to enhance or alter work activity performance, but this should be done only in a well-organized environment such as a change control process.

Forecasting Updates

Another important use of information from the results of change control is in updating forecasts of the schedule and budget that need to be communicated to individuals within the organization who require this information. Because the first order of business is to monitor work activity performance to ensure it stays on the current forecast (baseline), if change control is implemented, the project manager has to make adjustments to the forecast so that other departments, project staff, procurements, and other individuals external to the organization know there have been changes that may influence a work activity or other work packages in the project life cycle.

At the beginning of a project, the forecast represents what is initially planned in the schedule, the budget, and the expected quality of the deliverable to accomplish the project objective. As the project develops and the work activities are completed, if a change has to be implemented, it needs to be documented in the forecast and recommunicated to all individuals requiring the forecast and knowledge of what has changed. The project manager is responsible for not only maintaining the project management plan, which includes the schedule, budget, and analysis of information gathered during the monitoring process, but also managing the change control process to properly document and communicate the effects of change, which include updated forecasts.

Project Management Plan Updates

The project manager also has responsibility for developing the project management plan and documenting changes that have been made so that the project management plan stays current with actual work activity characteristics or parameters. The project manager also must update the schedule and budget in the project management plan to reflect how modifications have affected the plan. The project management plan is also the primary compilation of all other subsidiary plans that the project manager utilizes during the project life cycle and is archived with the understanding that it represents the forecast of information documenting actual performance.

Organizational Process Updates

If the change control process results in organizational processes that have been modified for the purpose of controlling project work activities, it is important that the project manager determine whether the process needs to be altered permanently or a temporary modification was needed for the purpose of meeting a specific work activity requirement. If an organizational process is used for project work activities, it is important the project manager understand that the process be performed as it was designed to ensure both the process and user will not be a root cause of failure. If modifications have to be made to a process because of a specific unique requirement for a work activity parameter, the modification should be documented as temporary. If it is determined through root cause analysis that the process itself is at fault by design, an organizational update is required regarding the correction of the process.

That update may involve other departments or individuals who need to sign off on this type of change. It's also important for the project manager to understand that projects are a temporary endeavor to create a unique deliverable and that there is typically nothing temporary or unique about standard organizational processes. Therefore, the warning to project managers is to be careful about permanent changes to organizational processes when used on project work activities.

Review Questions

1. Explain why change control is considered a process.
2. Discuss how the critical chain method is used to control a project.
3. Explain how schedule crashing works.
4. Discuss how cost control can be accomplished using contracts.
5. Are regulatory inspections considered part of quality control? If so, why?
6. Discuss why updating project forecasts would be necessary.

Applications Exercise

JP Phentar Construction: Case Study (Chapter 11)

Apply the concepts described in this chapter to the case study:

1. Determine the most effective schedule, cost, and quality control tools that could be utilized on this project. List the work activity and what tool would best control that activity.
2. Could the critical chain method be implemented on this project? If so, at what parts?
3. Could schedule fast tracking be used on this project? If so, explain how.
4. Are any quality control inspections or regulatory inspections required on this project?

Bibliography

Bell, Arthur H., and Dayle M. Smith. *Management Communication.* Hoboken, NJ: Wiley, 2006.

Bender, Michael B. *A Manager's Guide to Project Management: Learn How to Apply Best Practices.* Upper Saddle River, NJ: Pearson Education Inc. Publishing as FT Press, 2010.

Evans, James R., William M. Lindsay, and James R. Evans. *Managing for Quality and Performance Excellence.* Mason, OH: Thomson/South-Western, 2008.

Fleming, Quentin W., and Joel M. Koppelman. *Earned Value: Project Management.* 4th ed. Newtown Square, PA: Project Management Institute Inc., 2010.

Gido, Jack, and James P. Clements. *Successful Project Management.* 5th ed. Mason, OH: South-Western Cengage Learning, 2012.

Gray, Clifford F., and Erik W. Larson. *Project Management: The Managerial Process.* Boston: McGraw-Hill/Irwin, 2006.

Griffin, Ricky W. *Management.* Boston: Houghton Mifflin, 2005.

Heizer, Jay, and Barry Render. *Operations Management: Flexible Version.* Upper Saddle River, NJ: Pearson/Prentice Hall, 2007.

Heizer, Jay H., and Barry Render. *Principles of Operations Management.* Upper Saddle River, NJ: Pearson/Prentice Hall, 2004.

Hiegel, James, Roderick James, and Frank Cesario. *Projects, Programs, and Project Teams: Advanced Program Management.* Hoboken, NJ: Wiley Custom Services, 2006.

Kerzner, Harold. *Project Management: A Systems Approach to Planning, Scheduling, and Controlling.* 8th ed. Hoboken, NJ: Wiley, 2003.

Kuehn, Ursula, PMP, EVP. *Integrated Cost and Schedule Control in Project Management.* 2nd ed. Vienna, VA: Management Concepts Inc., 2011.

Lussier, Robert N., and Christopher F. Achua. *Leadership: Theory, Application, Skill Development*. Mason, OH: Thomson/SouthWestern, 2007.

Martino, Joseph P. *Research and Development Project Selection*. New York: Wiley, 1995.

Morris, Peter W. G., and Jeffrey K. Pinto. *The Wiley Guide to Project Control*. Hoboken, NJ: John Wiley & Sons, Inc., 2007.

Morris, Peter W. G., and Jeffrey K. Pinto. *The Wiley Guide to Project Program & Portfolio Management*. Hoboken, NJ: John Wiley & Sons, Inc., 2007.

Nicholas, John M., and Herman Steyn. *Project Management for Business, Engineering, and Technology: Principles and Practice*. Amsterdam: Elsevier Butterworth Heinemann, 2008.

Pinkerton, William J. *Project Management: Achieving Project Bottom-line Success*. Hightstown, NJ: The McGraw-Hill Companies, Inc., 2003.

Pinto, Jeffrey K. *Project Management: Achieving Competitive Advantage*. 3rd ed. Upper Saddle River, NJ: Pearson Education Inc., 2013.

Project Management Institute. *A Guide to the Project Management Body of Knowledge* (PMBOK® Guide), 5th ed. Newtown Square, PA: Project Management Institute, 2013.

Schermerhorn, John R., Richard Osborn, and James G. Hunt. *Organizational Behavior*. New York: Wiley, 2005.

Souder, William. E. *Project Selection and Economic Appraisal*. New York: Van Nostrand Reinhold, 1984.

Vaidyanathan, Ganesh. *Project Management: Process, Technology, and Practice*. Upper Saddle River, NJ: Pearson Education Inc., 2013.

Verma, Vijay K. *Organizing Projects for Success*. Upper Darby, PA: Project Management Institute, 1995.

Wilson, Randal. *The Operations Manager's Toolbox: Using the Best Project Management Techniques to Improve Processes and Maximize Efficiency*. Upper Saddle River, NJ: FT Press, 2013.

Index

Numbers

50/50 rule, 170

A

AC (actual cost), 230
accuracy of cost estimation data, 189
activity analysis, 80-88
    activity information checklist, 80-82
    activity organization, 83-84
    information gathering, 82-83
    in precedence diagramming method (PDM), 113-114
    in work breakdown structure (WBS), 85-88
activity contingency estimating, 145
activity decomposition decision tree, 71-72
activity definition, 79-80
    activity analysis, 80-88
        activity information checklist, 80-82
        activity organization, 83-84
        information gathering, 82-83
        in work breakdown structure (WBS), 85-88
    responsibility assignment, 88-91
        direct/indirect involvement, 88-89
        matrices, 90-91
    work authorization, 91-95
        defining, 93-95
        by organizational structure, 92-93
activity dependency matrix, 106, 166-167
activity disposition structure, 163
activity duration estimating, 139-140
    categories of, 151-152
    constraints, 146-151
    importance of, 152-153
    methods for, 141-146
    milestones, 153-154
    program management, 154-155
activity hierarchy structure, 164
activity information checklist, 80-82
activity level (organizing work activities), 84
activity sequencing, 97-98
    defining dependencies, 102-104
    information gathering, 98-101
        diagramming methods, 99-100
        terminology, 100-101
        type of information required, 99
    precedence diagramming method (PDM), 104-116
        activity analysis, 113-114
        activity dependency relationships, 105-106
        activity-on-node (AON) diagramming technique, 106-107
        critical path determination, 110-113
        float/slack calculation, 114-116
        nodes in, 107-108
        path types, 108-110
activity-on-arrow (AOA) diagramming technique, 100
activity-on-node (AON) diagramming technique, 100, 106-107
actual cost (AC), 230
ADM (arrow diagramming method), 100
administrative costs, 190
alternatives analysis (resource estimating), 131
analogous budgeting, 207-208
analogous cost estimating, 193-194
analogous estimating (activity duration), 141
analysis
    activity analysis, 80-88
        activity information checklist, 80-82
        activity organization, 83-84

information gathering, 82-83
in precedence diagramming method (PDM), 113-114
in work breakdown structure (WBS), 85-88
activity duration estimating
  reserve analysis, 145
  scenario analysis, 150-151
alternatives analysis (resource estimating), 131
change validation analysis, 238
cost estimation, reserve analysis, 197-198
earned value analysis (EVA), 229-232
fault tree analysis (FTA), 234
information analysis tools for monitoring projects, 223-232
make-or-buy analysis, 212
milestones, 224-225
project S-curve analysis, 223-224
root cause analysis, 233
schedule analysis, 171-180
  resource leveling, 174-175
  resource loading, 172-174
  scenario analysis, 178
  schedule reduction analysis, 175-178
  schedule variance analysis, 179-180
trend analysis, 225-227
variance analysis (schedules), 179-180
AOA (activity-on-arrow) diagramming technique, 100
AON (activity-on-node) diagramming technique, 100, 106-107
arrow diagramming method (ADM), 100
artifacts, project charter as, 54-55
authority, 91-95
  defining, 93-95
  by organizational structure, 92-93
availability of resources
  as constraint, 125
  project selection process, 36

B

BAC (budget at completion), 202-203, 230
backward pass
  defined, 100
  in precedence diagramming method (PDM), 114
balancing
  costs, 3
  resources, 2-3
baselines, creating, 227

beta distribution method (three-point estimating), 143-144, 194-196
bottom-up budgeting, 206-207
bottom-up constraints, 147-148
bottom-up cost estimating, 196-197
bubble diagrams (project selection process), 48
budget at completion (BAC), 202-203, 230
budget contingency planning, 210
budget development, 201
  constraints, 208-210
  cost of quality, 210-212
  methods for, 205-208
  purpose of, 202-204
buffers in critical chain method (CCM), 169-171, 249-250
burst activities
  defined, 100
  in precedence diagramming method (PDM), 109

C

capability
  in project selection models, 47
  as resource constraint, 124-125
capital equipment resources, 120
CCM (critical chain method), 169-171, 249-250
change control process, 69-70, 242-247
  activity duration estimating, 148
  authorization for, 94
  managing, 261
  in schedule development, 162
change validation analysis, 238
charter for project, 52-55, 64
check charts, 222-223
closure phase, 23
collecting data. *See* information gathering
communication
  in change control process, 245
  of schedule, 181-182
communications management plan, 61
conceptual phase, 23
  initiating process, 27-28
    origination of project, 28-31
    project charter, 52-55
    selecting projects, 34-51
    stakeholders, 31-34
constraints, 99
  activity duration estimating, 146-151
  budget development, 208-210
  for cost estimation, 191-192
  resource-constrained projects, 133

on resources, 122-125
theory of constraints (TOC), 167-169
time-constrained projects, 132
contingency control, 248
contingency estimating
activity duration estimating, 145
cost estimating, 197-198
contingency plans
authorization for, 94
budget development, 210
contract negotiation authorization for, 94
contracted resources, 121
contracts in cost control, 256-257
control
change control process, 242-247
contingency control, 248
cost control, 253-257
defined, 3
manager's role, 4
monitoring versus, 241-242
quality control, 257-260
reporting versus managing, 4
results, 260-263
schedule control, 249-253
control charts, 225-227
corrective action requirements, 237
cost aggregation budgeting, 206-207
cost control, organizational structure and, 13. *See also* control
cost in project selection models, 47
cost management plan, 60
budget development, 201
constraints, 208-210
cost of quality, 210-212
methods for, 205-208
purpose of, 202-204
estimating costs, 185
constraints, 191-192
information gathering, 186-190
methods for, 192-198
cost of quality, 210-212
cost performance index (CPI), 230
cost variance (CV), 230, 238
costs
balancing, 3
monitoring. *See* monitoring projects
CPI (cost performance index), 230
CPM (critical path method), 99-100, 166
critical chain method (CCM), 169-171, 249-250
critical path
defined, 100
determining, 110-113

critical path method (CPM), 99-100, 166
culture of organization, influence on projects, 6-7
customer specifications
activity duration estimating, 148
cost estimation, 192
requirements collection, 64
in schedule development, 162
schedule development, 160
customer-based programs, 42-43
CV (cost variance), 230, 238

D

data collection. *See* information gathering
decimal breakdown methodology, 74-75
deliverables, 63
Delphi method (resource estimating), 130
dependencies
in activity sequencing, 102-104
determining relationships, 105-106
design reviews, 259-260
determinate estimating (resource estimation), 130-131
diagramming methods for activity sequencing, 99-100, 104-116
direct costs, 190
direct project resources, 121
discretionary dependencies, 104
documentation of schedule, 180-182
dropped baton, 169
duration estimating. *See* activity duration estimating

E

EAC (estimate at completion), 231
early finish date (EF), 100
early start date (ES), 100
earned value analysis (EVA), 229-232
earned value (EV), 230
ease of use in project selection models, 47
estimate at completion (EAC), 231
estimate to completion (ETC), 231, 238
estimating
activity duration, 139-140
categories of, 151-152
constraints, 146-151
importance of, 152-153
methods for, 141-146
milestones, 153-154
program management, 154-155

costs, 185
    constraints, 191-192
    information gathering, 186-190
    methods for, 192-198
resources, 117-119
    constraints, 122-125
    methods for, 128-136
    requirements, 126-128
    types of resources, 119-122
ETC (estimate to completion), 231, 238
EV (earned value), 230
EVA (earned value analysis), 229-232
evaluating human resources, 117-119
    constraints, 122-125
    methods for, 128-136
    requirements, 126-128
    types of resources, 119-122
events, 100
execution phase, 23
    controlling, 241-242
        change control process, 242-247
        contingency control, 248
        cost control, 253-257
        quality control, 257-260
        results, 260-263
        schedule control, 249-253
    monitoring, 217-218
        information analysis tools, 223-232
        information gathering tools, 221-223
        integrated monitoring, 218-221
        results monitoring, 235-238
        sources of information, 221
        troubleshooting tools, 233-234
expectations of stakeholders, 33-34
expected costs, 203
external dependencies, 104
external factors in schedule development, 160
external projects, initiating process, 30-31
external requirements, 63
external resources, 121

F

facilities resources, 120
fast tracking, 177-178, 252-253
fault tree analysis (FTA), 234
50/50 rule, 170
financial models (project selection process), 49-51

financial resources. *See also* cost management plan
    defined, 120
    project selection process, 36-37
finish-to-finish (FF) relationship, 105
finish-to-start (FS) relationship, 105
fixed-price contracts, 256
fixed-price incentive fee contracts, 257
flexibility in project selection models, 47
float/slack
    calculating, 114-116
    defined, 100
forecasting
    adjustment requirements, 237-238
    updates, 261-262
forward pass
    defined, 101
    in precedence diagramming method (PDM), 113
FTA (fault tree analysis), 234
functional organizations
    defined, 7-8
    project selection process, 37-38
    work authorization, 92
funding limit reconciliation, 209

H

historical data
    budget development, 207-208
    requirements collection, 65
in-house technology (project selection process), 35
human resources
    defined, 119
    evaluating, 117-119
        constraints, 122-125
        methods for, 128-136
        requirements, 126-128
        types of resources, 119-122
    management plan, 61
    project selection process, 35
    responsibility assignment, 88-91
        direct/indirect involvement, 88-89
        matrices, 90-91
    work authorization, 91-95
        defining, 93-95
        by organizational structure, 92-93

I

identified risks, 99
implementation in change control process, 244-245
independent projects, selection process, 46
indirect costs, 190
indirect labor costs, 190
indirect materials cost, 190
indirect project resources, 121
information analysis tools for monitoring projects, 223-232
information gathering
    for activity duration estimating, 151
    for activity sequencing, 98-101
        diagramming methods, 99-100
        terminology, 100-101
        type of information required, 99
    for cost control, 254-255
    for cost estimation, 186-190
        accuracy and reliability, 189
        cost requirements, 186-187
        direct versus indirect costs, 189-190
        sources, 187-189
    for monitoring projects, 221-223
    for schedule control, 249
    for schedule development, 158-160
    for work activities, 82-83
information technology resources, 120
initiating process, 27-28
    origination of project, 28-31
    project charter, 52-55
    selecting projects, 34-51
        independent projects, 46
        models and methodologies, 47-51
        organizational constraints, 34-38
        in organizational strategy, 40-42
        for portfolios and programs, 42-46
        project management constraints, 38-40
    stakeholders, 31-34
inspections
    quality inspections, 258-259
    regulatory inspections, 259
integrated change control, 246-247
integrated monitoring, 218-221
intended functionality (measure of quality), 257
internal projects, initiating process, 29-30
internal requirements, 63

L

labels. *See* nodes
late finish date (LF), 101
late start date (LS), 101
leadership, influence on projects, 5-6
life cycle of projects, 23-25. *See also* closure phase; conceptual phase; execution phase; planning phase

M

make-or-buy analysis, 212
managers
    project selection process, 35-36
    role in project control, 4
managing
    change control process, 261
    reporting versus, 4
mandatory dependencies, 104
materials and workmanship standards (measure of quality), 258
materials resources, 120
matrix organizations
    defined, 9
    project selection process, 37-38
    work authorization, 93
measurement in change control process, 245-246
meetings, status, 221-222
merge activities
    defined, 101
    in precedence diagramming method (PDM), 109
methodologies (project selection process), 47-51
    qualitative models, 48
    quantitative models, 49-51
Microsoft Excel
    for schedule documentation, 181
    in work breakdown structure (WBS), 86
Microsoft Project
    baseline creation, 227
    for schedule documentation, 181
    Tracking Gantt charts, 228
    in work breakdown structure (WBS), 86-87
milestones
    analysis, 224-225
    reporting status, 204
    in schedule development, 153-154

models (project selection process), 47-51
    qualitative models, 48
    quantitative models, 49-51
monitoring projects, 217-218
    controlling projects versus, 241-242
    information analysis tools, 223-232
    information gathering tools, 221-223
    integrated monitoring, 218-221
    results monitoring, 235-238
    sources of information, 221
    troubleshooting tools, 233-234
multiple critical paths, 112-113
multitasking, 170

N

net present value (NPV), calculating, 50-51
network diagrams. *See also* diagramming methods
    in activity duration estimating, 152
    defined, 101
    in schedule development, 165-167
new risk assessment, 236-237
nodes
    defined, 101
    in precedence diagramming method (PDM), 107-108
NPV (net present value), calculating, 50-51

O

organizational constraints
    cost estimation, 191
    project selection process, 34-38
    resource estimating, 123
    schedule development, 160
organizational division-based programs, 44-45
organizational influences on projects, 4-9
    culture, 6-7
    leadership, 5-6
    structure, 7-9
        cost control and, 13
        explained, 14
        profit centers versus support functions, 14-15
        project selection process, 37-38
        work authorization by, 92-93
organizational needs, projects/programs/portfolios and, 18
organizational process updates, 262-263
organizational resource management, 127-128
organizational strategy (project selection process), 40-42
organizing work activities, 83-84
origination of projects, 28-31
outsource contracting in cost of quality, 211-212
overhead expenses, 190

P

parallel activities
    defined, 101
    in precedence diagramming method (PDM), 109
parametric cost estimating, 194
parametric estimating (activity duration), 141-142
Parkinson's Law, 169
paths
    defined, 101
    in precedence diagramming method (PDM), 108-110
payback period (project selection process), 50
PDM (precedence diagramming method), 100, 104-116
    activity analysis, 113-114
    activity dependency relationships, 105-106
    activity-on-node (AON) diagramming technique, 106-107
    critical path determination, 110-113
    float/slack calculation, 114-116
    nodes in, 107-108
    path types, 108-110
    in schedule development, 166-167
performance reports, 235-236
PERT (program evaluation and review technique), 99-100
phases of projects, 23-25. *See also* closure phase; conceptual phase; execution phase; planning phase
planned value (PV), 229
planning phase, 23, 57-58
    cost management plan, 60. *See also* cost management plan
    project management plan, 58-62
    requirements collection, 63-66
        defining requirements, 63-64
        management plan, 66
        resources, 64-65
    schedule management plan, 60. *See also* schedule management plan

INDEX **273**

scope definition, 66-70
    change control process, 69-70
    product scope, 67
    project scope, 67
    project scope statement, 68
    responsibility for, 67-68
work breakdown structure (WBS), 70-75
portfolio management
    defined, 18
    importance of, 21-22
    project selection process, 39-40
    resource estimating, 128-129
    responsibilities of, 21
portfolios
    organizational needs and, 18
    project selection process, 42-46
    projects and programs versus, 15-17
precedence diagramming method (PDM), 100, 104-116
    activity analysis, 113-114
    activity dependency relationships, 105-106
    activity-on-node (AON) diagramming technique, 106-107
    critical path determination, 110-113
    float/slack calculation, 114-116
    nodes in, 107-108
    path types, 108-110
    in schedule development, 166-167
predecessor constraints (activity duration estimating), 149-150
predecessor requirements, 74
predecessors
    creating relationships, 102-104
    defined, 101, 105
process, project charter as, 54-55
procrastination, 170
procurement management plan, 61
procurements
    in cost control, 255-256
    in cost of quality, 211
product scope, 67
product-based programs, 43-44
profit centers, support functions versus, 14-15
program evaluation and review technique (PERT), 99-100
program management
    activity duration estimating, 154-155
    defined, 18
    importance of, 21-22
    project selection process, 39-40
    resource estimating, 129
    responsibilities of, 19-21

programs
    organizational needs and, 18
    project selection process, 42-46
    projects and portfolios versus, 15-17
project budget baseline, 203
project charter, 52-55
    requirements collection, 64
    schedule development, 159
project constraints
    cost estimation, 191
    resource estimating, 123-124
project contingency estimating, 145
project deliverables, 101
project level (organizing work activities), 84
project management
    activity duration estimating, 154-155
    defined, 18
    importance of, 21-22, 57-58
    plans, 58-62
        defined, 58
        structure, 59-62
        updates, 262
        usage, 62
    project selection process, constraints in, 38-40
    resource estimating, 129
    responsibilities of, 18-19
project milestones
    analysis, 224-225
    reporting status, 204
    in schedule development, 153-154
project monitoring information systems, 218-221
project plans, 59
project resource requirements, 126-127
project scope, 67, 161. *See also* scope
project scope statement, 68, 159
project S-curve analysis, 223-224
projectized organizations
    defined, 8-9
    project selection process, 37-38
    work authorization, 92
projects
    control
        defined, 3
        manager's role, 4
        reporting versus managing, 4
    costs, balancing, 3
    defined, 1-2
    execution phase
        controlling, 241-242. See also control
        monitoring, 217-218. See also monitoring
            projects

initiating process, 27-28
    origination of project, 28-31
    project charter, 52-55
    selecting projects, 34-51
    stakeholders, 31-34
life-cycle phases, 23-25
organizational influences, 4-9
    culture, 6-7
    leadership, 5-6
    structure, 7-9
organizational needs and, 18
planning phase, 57-58
    cost management plan, 60. See also cost management plan
    project management plan, 58-62
    requirements collection, 63-66
    schedule management plan, 60. See also schedule management plan
    scope definition, 66-70
    work breakdown structure (WBS), 70-75
programs and portfolios versus, 15-17
requirements, 63
resources, balancing, 2-3
structuring, 13-14
proposals
    in change control process, 243-244
    requesting, 30-31
published data estimating (resource estimation), 131
PV (planned value), 229

Q

qualitative project selection, 41-42, 48
quality
    controlling, 257-260
    cost of, 210-212
quality inspections, 258-259
quality management plan, 60
quantitative project selection, 42, 49-51
quotes, requesting, 31

R

RACI (responsibility, accountability, consultative, and informative) matrix, 90-91
RAM (responsibility assignment matrix), 90-91
realism in project selection models, 47
reduction in schedule duration, 175-178
regulatory inspections, 259
reliability of cost estimation data, 189

reporting
    managing versus, 4
    project control, 260-261
    project status, 204
    work performance, 235-236
request for proposal (RFP), 30-31
request for quote (RFQ), 31
requirements
    collecting, 63-66
    corrective action requirements, 237
    cost estimation, 186-187
    defined, 63-64
    forecasting adjustment requirements, 237-238
    management plan, 66
    predecessor/successor, 74
    resources
        determining, 126-128
        information gathering, 64-65
        resource requirements plan, 135-136
        in schedule development, 161
reserve analysis
    activity duration estimating, 145
    cost estimation, 197-198
resource assessment, 90
resource requirements plan, 135-136
resource-constrained projects, 133
resources
    balancing, 2-3
    bottlenecks, 170
    constraints, 124-125
    estimating, 117-119
        constraints, 122-125
        methods for, 128-136
        requirements, 126-128
        types of resources, 119-122
    leveling, 134, 174-175, 250-251
    loading, 133-134, 172-174
    outsourcing, 211-212
    project selection process, 36
    requirements, 64-65, 161
responsibilities
    assigning, 88-91
        direct/indirect involvement, 88-89
        responsibility assignment matrices, 90-91
    authority for, 91-95
        defining, 93-95
        by organizational structure, 92-93
    of portfolio management, 21
    of program management, 19-21
    of project management, 18-19

responsibility, accountability, consultative, and informative (RACI) matrix, 90-91
responsibility assignment matrix (RAM), 90-91
results
    of controlling projects, 260-263
    of monitoring projects, 235-238
return on investment (ROI), calculating, 51
RFP (request for proposal), 30-31
RFQ (request for quote), 31
risk contingency, authorization for, 94
risk management plan, 61
risks
    identifying, 99
    new risk assessment, 236-237
ROI (return on investment), calculating, 51
ROME (rough order of magnitude estimating), 193
root cause analysis, 233
rough order of magnitude estimating (ROME), 193

S

scenario analysis
    activity duration estimating, 150-151
    schedule development, 178
schedule analysis, 171-180
    resource leveling, 174-175
    resource loading, 172-174
    scenario analysis, 178
    schedule reduction analysis, 175-178
    schedule variance analysis, 179-180
schedule crashing, 176-177, 252
schedule development, 157-158
    information gathering, 158-160
        customer requirements, 162
        project scope, 161
        resource requirements, 161
    schedule analysis, 171-180
        resource leveling, 174-175
        resource loading, 172-174
        scenario analysis, 178
        schedule reduction analysis, 175-178
        schedule variance analysis, 179-180
    schedule documentation, 180-182
    structuring schedules, 162-171
        activity disposition structure, 163
        activity hierarchy structure, 164
        critical chain method (CCM), 169-171
        network diagrams, 165-167
        theory of constraints (TOC), 167-169

schedule management plan, 60
    activity definition, 79-80
        activity analysis, 80-88
        responsibility assignment, 88-91
        work authorization, 91-95
    activity duration estimating, 139-140
        categories of, 151-152
        constraints, 146-151
        importance of, 152-153
        methods for, 141-146
        milestones, 153-154
        program management, 154-155
    activity sequencing, 97-98
        defining dependencies, 102-104
        information gathering, 98-101
        precedence diagramming method (PDM), 104-116
    resource estimating, 117-119
        constraints, 122-125
        methods for, 128-136
        requirements, 126-128
        types of resources, 119-122
    schedule development, 157-158
        customer requirements, 162
        information gathering, 158-160
        project scope, 161
        resource requirements, 161
        schedule analysis, 171-180
        schedule documentation, 180-182
        structuring schedules, 162-171
schedule performance index (SPI), 230
schedule reduction analysis, 175-178
schedule variance analysis, 179-180
schedule variance (SV), 230, 238
scheduling
    authorization for, 93
    controlling. See control
    monitoring. See monitoring projects
scope
    change control process, 69-70
    defining, 66-70
        product scope, 67
        project scope, 67
        responsibility for, 67-68
    project scope statement, 68
    in schedule development, 161
scope creep, 69, 148, 187
scope management plan, 60
scoring model (project selection process), 48
selecting projects, 34-51
    independent projects, 46
    models and methodologies, 47-51

organizational constraints, 34-38
   in organizational strategy, 40-42
   for portfolios and programs, 42-46
   project management constraints, 38-40
self-protection, 169
sequencing activities, 97-98
   defining dependencies, 102-104
   information gathering, 98-101
      diagramming methods, 99-100
      terminology, 100-101
      type of information required, 99
   precedence diagramming method (PDM), 104-116
      activity analysis, 113-114
      activity dependency relationships, 105-106
      activity-on-node (AON) diagramming technique, 106-107
      critical path determination, 110-113
      float/slack calculation, 114-116
      nodes in, 107-108
      path types, 108-110
serial activities
   defined, 101
   in precedence diagramming method (PDM), 108
SMEs (subject matter experts)
   activity duration estimating, 146
   cost estimation, 193
   monitoring projects, 222
   requirements collection, 65
software tools for schedule documentation, 181
SOW (statement of work), 64
spending, authorization for, 93
SPI (schedule performance index), 230
stakeholder management plan, 62, 65
stakeholder register, 65
stakeholders, 31-34
   expectations, 33-34
   managing, 31-33
   project charter, 52
   requirements, 63
start-to-finish (SF) relationship, 105
start-to-start (SS) relationship, 105
statement of work (SOW), 64
status meetings, 221-222
structure of organization
   cost control and, 13
   explained, 14
   influence on projects, 7-9

profit centers versus support functions, 14-15
   project selection process, 37-38
   work authorization by, 92-93
structuring
   project charters, 52-53
   project management plans, 59-62
   projects, 13-14
   schedules, 162-171
      activity disposition structure, 163
      activity hierarchy structure, 164
      critical chain method (CCM), 169-171
      network diagrams, 165-167
      theory of constraints (TOC), 167-169
student syndrome, **170**
subject matter experts (SMEs)
   activity duration estimating, 146
   cost estimation, 193
   monitoring projects, 222
   requirements collection, 65
successor constraints (activity duration estimating), 149-150
successor requirements, 74
successors
   creating relationships, 102-104
   defined, 101, 105
support functions, profit centers versus, 14-15
SV (schedule variance), 230, 238

T

task-on-arrow (TOA) diagramming technique, 100
terminology for activity sequencing, 100-101
theory of constraints (TOC), 167-169
three-point estimating
   activity duration estimating, 142-145
   cost estimating, 194-196
time and materials contracts, 257
time value of money (project selection process), 49-50
time-constrained projects, 132
time-phased budgeting, 207
TOA (task-on-arrow) diagramming technique, 100
TOC (theory of constraints), 167-169
top-down budgeting, 205-206
top-down constraints, 147-148
top-down cost estimating, 196-197
Tracking Gantt charts, 228
trend analysis, 225-227

triangular distribution method (three-point estimating), 143, 194-196
triple constraint
    activity duration estimating, 146-147
    budget development, 203-204
    controlling projects, 247
    in project constraints, 124
    quality costs, 210-212
troubleshooting tools for monitoring projects, 233-234

U

updates
    forecasting, 261-262
    organizational process, 262-263
    project management plan, 262

V

variance analysis (schedules), 179-180
variance at completion (VAC), 231

W

WBS (work breakdown structure), 70-75
    activity definition in, 85-88
    schedule development, 160
work activities
    activity definition, 79-80
        activity analysis, 80-88
        responsibility assignment, 88-91
        work authorization, 91-95
    activity duration estimating, 139-140
        categories of, 151-152
        constraints, 146-151
        importance of, 152-153
        methods for, 141-146
        milestones, 153-154
        program management, 154-155
    activity sequencing, 97-98
        defining dependencies, 102-104
        information gathering, 98-101
        precedence diagramming method (PDM), 104-116
    in schedule development, 160
    variance, 179-180
work authorization, 91-95
    defining, 93-95
    by organizational structure, 92-93

work breakdown structure (WBS), 70-75
    activity definition in, 85-88
    schedule development, 160
work package activities, defined, 101
work performance reports, 235-236